# Ecotourism, second edition

As arguably the largest and fastest-growing industry, the potential impacts of tourism are considerable. Devised by social and natural scientists, ecotourism can be an effective way to safeguard local communities and prevent new destruction of the natural world. Responding to increased interest and global competition, the tourism industry has not been slow to appropriate the term to describe a host of different experiences, markedly different from its origins.

The revised and updated edition of *Ecotourism* provides a broad introduction, including:

- The relationship of ecotourism to the broader tourism literature;
- History, definition and typologies of ecotourism;
- Social and ecological impacts of tourism;
- Role of tourism policy, regulation, certification, accreditation and professionalism;
- Economics, marketing and management of ecotourism;
- Ecotourism programme planning;
- Ecotourism and international development and the role of community development;
- The role of ethics.

The second edition incorporates new material on eco-labelling, environmental management and guiding. A new chapter has been added on programme planning and coverage of environmental impacts, and community-based management has been strengthened. The book also incorporates new developments stemming from the 2002 World Ecotourism meeting in Quebec. New case studies and examples have been added throughout, as well as annotated further reading.

**David A. Fennell**, PhD, is based in the Department of Recreation and Leisure Studies, Brock University, St Catherines, Ontario, Canada.

# Ecotourism

## An introduction

## Second edition

David A. Fennell

Routledge
Taylor & Francis Group

LONDON AND NEW YORK

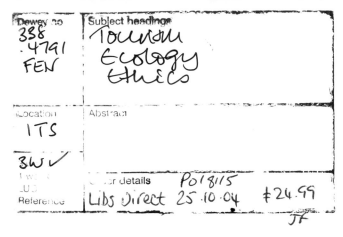
First published 1999 by Routledge
11 New Fetter Lane, London EC4P 4EE

Simultaneously published in the USA and Canada
by Routledge
29 West 35th Street, New York, NY 10001

Second edition first published 2003

*Routledge is an imprint of the Taylor & Francis Group*

© 1999, 2003 David A. Fennell

Typeset in Times by Keystroke, Jacaranda Lodge, Wolverhampton
Printed and bound in Great Britain by Bell & Bain Ltd, Glasgow

*British Library Cataloguing in Publication Data*
A catalogue record for this book is available from the British Library

*Library of Congress Cataloging in Publication Data*
Fennell, David A., 1963–
    Ecotourism: an introduction / David A. Fennell. — 2nd ed.
        p. cm.
    Includes bibliographical references (p. ).
    1. Ecotourism.  I.  Title.
    G156.5.E26F465 2003
    338.4′791—dc21                                    2002155143

ISBN 0–415–30364–8 (hbk)
ISBN 0–415–30365–6 (pbk)

To my family

# Contents

# Plates

# Figures

# Tables

# Case studies

# **Preface**

This book came to be written for three main reasons. The first of these was to address what might be considered 'inconsistencies' in the philosophical basis of ecotourism, and the development and implementation of ecotourism products in a wide variety of destinations. For example, in a sobering account of her travel experience in the Peruvian rainforest, Arlen (1995) writes that ecotourism has reached a critical juncture in its evolution. She speaks graphically of instances where tourists endured swimming in water with human waste; guides capturing sloths and caiman for tourists to photograph; raw sewage openly dumped into the ocean; mother cheetahs killing their cubs to avoid the harassment of cheetah-chasing tourists; and an ecotourism industry under-regulated with little hope for enforcement. Similar experiences have been recorded by other writers including Farquharson (1992), who argues that ecotourism is a dream that has been severely diluted. She writes that whereas birding once prevailed, ecotourism has fallen into the clutches of many of the mega-resorts like Cancún: The word [ecotourism] changes color like a chameleon. What began as a concept designed by ecologists to actively prevent the destruction of the environment has become a marketing term for tourism developers who want to publicize clean beaches, fish-filled seas and a bit of culture for when the sunburn begins to hurt (Farquharson 1992: 8). These scenarios appear to be worlds apart from the evolution of ecotourism in the not too distant past, where, as outlined by Farquharson, it was seen as a haven for birdwatchers and scientists alike. Clearly ecotourism is a thriving economic enterprise in both developed and less developed countries around the world. However, while scientists occupy one end of the ecotourism continuum, in other cases this form of tourism has come to represent a completely different type of experience, with the industry clamouring to take advantage of a larger and softer market of ecotourists, as a result of increased interest and competition. According to some (Budowski in Arlen, 1995), ecotravellers at this softer end of the continuum have learned to expect a type of experience much like what one might get in Hawaii or Cancún. The industry involvement is just one of many facets of the ecotourism industry discussed in this book. Others include government involvement in ecotourism, aboriginal interests, partnership and training, tourist demand, structural differences between developed and developing countries, policy and regulation, ethics and responsibility, and so on.

Second, the book was undertaken to demonstrate the fact that there is a vast amount of ecotourism material that is currently available in the literature. Literally hundreds of articles have been written on ecotourism – academic and non-academic – many of which have surfaced in the past seven or eight years. It was felt that an introductory book would address at least some of this literature in addition to many key issues related to the field.

Finally, a review of literature made it clear that few texts of this nature are currently available to help students in their understanding of the topic area. A significant body of literature cited in the book, especially in regard to tourism and recreation, falls outside the realm of ecotourism research. This is intentional, as it indirectly suggests that in many cases much valuable tourism research is neglected by ecotourism writers and researchers. Because of the infancy of ecotourism research there are many 'unknowns' that may be partially addressed by the general tourism literature and literature from other disciplines.

# Acknowledgements

A number of people have been instrumental in helping to shape the ideas found in this book. My initial interest in ecotourism was nurtured by Paul Eagles and Bryan Smale at the University of Waterloo, Canada. Both provided unique perspectives on recreation and tourism which enabled me to build an objective view of the relevance and place of ecotourism in society. Thanks are extended to R.W. Butler, whom I was fortunate to work with at the University of Western Ontario. Dick challenged me to broaden my understanding of tourism, beyond ecotourism, and its application to geographical phenomena. I wish to acknowledge my many colleagues around the world who continue to inspire all who are interested in this dynamic field. Curt Shroeder and Nicole Choptain deserve thanks for helping in the collection of information for the first edition; while Val Sheppard is acknowledged for her help with this second edition.

Outside of academia, and most importantly, I wish to acknowledge my family, who provided endless support and encouragement throughout the writing of this book (and in all other endeavours of my life). My parents and brothers continue to be strong influences in my life, as are my wife, son and daughter who make each day better than the one before it.

# **1** The nature of tourism

In this chapter the tourism system is discussed, including definitions of tourism and associated industry elements. Considerable attention is paid to attractions as fundamental elements of the tourist experience. Both mass tourism and alternative tourism paradigms are introduced as a means by which to overview the philosophical approaches to tourism development to the present day. Finally, much of the chapter is devoted to sustainable development and sustainable tourism, including sustainable tourism indicators, for the purpose of demonstrating the relevance of this form of development to the future of the tourism industry. This discussion will provide a backdrop from which to analyse ecotourism, which is detailed at length in Chapter 2.

## Defining tourism

As one of the world's largest industries, tourism is associated with many of the prime sectors of the world's economy. Any such phenomenon that is intricately interwoven into the fabric of life economically, socio-culturally, and environmentally and relies on primary, secondary, and tertiary levels of production and service, is difficult to define in simple terms. This difficulty is mirrored in a 1991 issue of *The Economist*:

> There is no accepted definition of what constitutes the [tourism] industry; any definition runs the risk of either overestimating or underestimating economic activity. At its simplest, the industry is one that gets people from their home to somewhere else (and back), and which provides lodging and food for them while they are away. But that does not get you far. For example, if all the sales of restaurants were counted as travel and tourism, the figure would be artificially inflated by sales to locals. But to exclude all restaurant sales would be just as misleading.

It is this complex integration within our socio-economic system (a critical absence of focus), according to Clawson and Knetsch (1966) and Mitchell (1984), that complicates efforts to define tourism. Tourism studies are often placed poles apart in terms of philosophical approach, methodological orientation, or intent of the investigation. A variety of tourism definitions, each with disciplinary attributes, reflect research initiatives corresponding to various fields. For example, tourism shares strong fundamental characteristics and theoretical foundations with the recreation and leisure studies field. According to Jansen-Verbeke and Dietvorst (1987) the terms 'leisure', 'recreation' and 'tourism' represent a type of loose, harmonious unity which focuses on the experiential and activity-based features that typify these terms. On the other hand, economic and technical/statistical definitions generally ignore the human experiential elements of the concept in favour of an approach based on the movement of people over political borders and the amount of money generated from this movement.

It is this relationship with other disciplines, e.g. psychology, sociology, anthropology, geography, economics, which seems to have defined the complexion of tourism. However, despite its strong reliance on such disciplines, some, including Leiper (1981), have advocated a move away in favour of a distinct tourism discipline. To Leiper the way in which we need to approach the tourism discipline should be built around the structure of the industry, which he considers as an open system of five elements interacting with broader environments: (1) a dynamic human element, (2) a generating region, (3) a transit region, (4) a destination region, and (5) the tourist industry. This definition is similar to one established by Mathieson and Wall (1982), who see tourism as comprising three basic elements: (1) a dynamic element, which involves travel to a selected destination; (2) a static element, which involves a stay at the destination; and (3) a consequential element, resulting from the above two, which is concerned with the effects on the economic, social, and physical subsystems with which the tourist is directly or indirectly in contact. Others, including Mill and Morrison, define tourism as a system of interrelated parts. The system is 'like a spider's web – touch one part of it and reverberations will be felt throughout' (Mill and Morrison 1985: xix). Included in their tourism system are four component parts, including Market (reaching the marketplace), Travel (the purchase of travel products), Destination (the shape of travel demand), and Marketing (the selling of travel).

In recognition of the difficulty in defining tourism, Smith (1990a) feels that it is more realistic to accept the existence of a number of different definitions, each designed to serve different purposes. This may in fact prove to be the most practical of approaches to follow. In this book, tourism is defined as the interrelated system that includes tourists and the associated services that are provided and utilised (facilities, attractions, transportation, and accommodation) to aid in their movement, while a tourist, as established by the World Tourism Organization, is defined as a person travelling for pleasure for a period of at least one night, but not more than one year for international tourists and six months for persons travelling in their own countries, with the main purpose of the visit being other than to engage in activities for remuneration in the place(s) visited.

## Tourism attractions

The tourism industry includes a number of key elements that tourists rely upon to achieve their general and specific goals and needs within a destination. Broadly categorised, they include facilities, accommodation, transportation, and attractions. Although an in-depth discussion of each is beyond the scope of this book, there is merit in elaborating upon the importance of tourism attractions as a fundamental element of the tourist experience. These may be loosely categorised as cultural (e.g., historical sites, museums), natural (e.g., parks, flora and fauna), events (e.g., festivals, religious events), recreation (e.g., golf, hiking), and entertainment (e.g., theme parks, cinemas), according to Goeldner *et al.* (2000). Past tourism research has tended to rely more on the understanding of attractions, and how they affect tourists, than of other components of the industry. As Gunn has suggested, 'they [attractions] represent the most important reasons for travel to destinations' (1972: 24).

MacCannell described tourism attractions as 'empirical relationships between a tourist, a site and a marker' (1989: 41). The tourist represents the human component, the site includes the actual destination or physical entity, and the marker represents some form of information that the tourist uses to identify and give meaning to a particular attraction. Lew (1987), however, took a different view, arguing that under the conditions of tourist–site–marker, virtually anything could become an attraction, including services and facilities. Lew chose to emphasise the objective and subjective characteristics of attractions by suggesting that researchers ought to be concerned with three main areas of the attraction:

- *Ideographic*. Describes the concrete uniqueness of a site. Sites are individually identified by name and usually associated with small regions. This is the most frequent form
  of attraction studied in tourism research.
- *Organisational*. The focus is not on the attractions themselves, but rather on their spatial
  capacity and temporal nature. Scale continua are based on the size of the area which
  the attraction encompasses.
- *Cognitive*. A place that fosters the feeling of being a tourist. Attractions are places
  that elicit feelings related to what Relph (1976) termed 'insider' 'outsider', and the
  authenticity of MacCannell's (1989) front and back regions.

Leiper (1990: 381) further added to the debate by adapting MacCannell's model into a
systems definition. He wrote that:

> A tourist attraction is a systematic arrangement of three elements: a person with touristic
> needs, a nucleus (any feature or characteristic of a place they might visit) and at least one
> marker (information about the nucleus).

The type of approach established by Leiper is also reflected in the efforts of Gunn (1972),
who has written at length on the importance of attractions in tourism research. Gunn
produced a model of tourist attractions that contained three separate zones, including (1) the
nuclei, or core of the attraction; (2) the inviolate belt, which is the space needed to set
the nuclei in a context; and (3) the zone of closure, which includes desirable tourism
infrastructure such as toilets and information. Gunn argued that an attraction missing one
of these zones will be incomplete and difficult to manage.

Some authors, including Pearce (1982), Gunn (1988) and Leiper (1990), have made
reference to the fact that attractions occur on various hierarchies of scale, from very specific
and small objects within a site to entire countries and continents. This scale variability
further complicates the analysis of attractions as both sites and regions. Consequently, there
exists a series of attraction cores and attraction peripheries, within different regions, between
regions, and from the perspective of the types of tourists who visit them. Spatially, and with
the influence of time, the number and type of attractions visited by tourists and tourist groups
may create a niche; a role certain types of tourists occupy within a vacation destination.
Through an analysis of space, time, and other behavioural factors, tourists can be fitted
into a typology based on their utilisation and travel between selected attractions. One could
make the assumption that tourist groups differ on the basis of the type of attractions they
choose to visit, and according to how much time they spend at them (see Fennell 1996).
The implications for the tourism industry are that often it must provide a broad range of
experiences for tourists interested in different aspects of a region. A specific destination
region, for example, may recognise the importance of providing a mix of touristic opportunities, from the very specific, to more general interest experiences for the tourists in search
of cultural and natural experiences, in urban, rural and back-country settings. ('Back' regions
are defined on p. 34.)

Attractions have also been referred to in past research as sedentary, physical entities of a
cultural or natural form (Gunn 1988). In their natural form, such attractions form the basis
for distinctive types of tourism which are based predominantly on aspects of the natural
world, such as wildlife tourism (see Reynolds and Braithwaite 2001), and ecotourism (see
Page and Dowling 2002). For example, to a birdwatcher individual species become
attractions of the most specific and most sought-after kind. A case in point is the annual return
of a single albatross at the Hermaness National Nature Reserve in Unst, Shetland, Scotland.
The arrival of this species prompts birdwatching tourists immediately to change their
plans in an effort to travel to Hermaness. The albatross has become a major attraction for

birder-tourists, while Hermaness, in a broader context, acts as a medium (attraction cluster) by which to present the attraction (bird). Natural attractions can be transitory in space and time, and this time may be measured for particular species in seconds, hours, days, weeks, months, seasons, or years. For tourists who travel with the prime reason to experience these transitory attractions, their movement is a source of both challenge and frustration.

## Mass and alternative tourism: competing paradigms

Tourism has been both lauded and denounced for its ability to develop and therefore transform regions into completely different settings. In the former case, tourism is seen to have provided the impetus for appropriate long-term development; in the latter the ecological and sociological disturbance to transformed regions can be overwhelming. While most of the documented cases of the negative impacts of tourism are in the developing world, the developed world is certainly not an exception. Young (1983), for example, documented the transformation of a small fishing farming community in Malta by graph-ically illustrating the extent to which tourism development – through an increasingly complex system of transportation, resort development, and social behaviour – overwhelms such areas over time.

These days we are more prone to vilify or characterise conventional mass tourism as a beast, a monstrosity which has few redeeming qualities for the destination region, their people and their natural resource base. Consequently, mass tourism has been criticised for the fact that it dominates tourism within a region owing to its non-local orientation, and the fact that very little money spent within the destination actually stays and generates more income. It is quite often the hotel or mega-resort that is the symbol of mass tourism's domination of a region, which are often created using non-local products, have little requirement for local food products, and are owned by metropolitan interests. Hotel marketing occurs on the basis of high volume, attracting as many people as possible, often over seasonal periods of time. The implications of this seasonality are such that local people are at times moved in and out of paid positions that are based solely on this volume of touristic traffic. Development exists as a means by which to concentrate people in very high densities, displacing local people from traditional subsistence-style livelihoods (as outlined by Young 1983) to ones that are subservience based. Finally, the attractions that lie in and around these massive developments are created and transformed to meet the expectations and demands of visitors. Emphasis is often on commercialisation of natural and cultural resources, and the result is a contrived and inauthentic representation of, for example, a cultural theme or event that has been eroded into a distant memory.

Admittedly the picture of mass tourism painted above is outlined to illustrate the point that the tourism industry has not always operated with the interests of local people and the resource base in mind. This was most emphatically articulated through much of the tourism research that emerged in the 1980s, which argued for a new, more socially and ecologically benign alternative to mass tourism development. According to Krippendorf (1982), the philosophy behind alternative tourism (AT) – forms of tourism that advocate an approach opposite to mass conventional tourism – was to ensure that tourism policies should no longer concentrate on economic and technical necessities alone, but rather emphasise the demand for an unspoiled environment and consideration of the needs of local people. This 'softer' approach places the natural and cultural resources at the forefront of planning and development, instead of as an afterthought. Also, as an inherent function, alternative forms of tourism provide the means for countries to eliminate outside influences, and to sanction projects themselves and to participate in their development – in essence, to win back the decision-making power in essential matters rather than conceding to outside people and institutions.

**Plate 1.1** Tourist development at Cancún, Mexico

AT is a generic term that encompasses a whole range of tourism strategies (e.g. 'appropriate', 'eco-', 'soft', 'responsible', 'people to people', 'controlled', 'small-scale', 'cottage', and 'green' tourism), all of which purport to offer a more benign alternative to conventional mass tourism in certain types of destinations (Conference Report 1990, cited in Weaver 1991). Dernoi (1981) illustrates that the advantages of AT will be felt in five ways:

1   There will be benefits for the individual or family: accommodation based in local homes will channel revenue directly to families. Also families will acquire managerial skills.
2   The local community will benefit: AT will generate direct revenue for community members, in addition to upgrading housing standards while avoiding huge public infrastructure expenses.
3   For the host country, AT will help avoid the leakage of tourism revenue outside the country. AT will also help prevent social tensions and may preserve local traditions.
4   For those in the industrialised generating country, AT is ideal for cost-conscious travellers or for people who prefer close contacts with locals.
5   There will be benefits for international relations: AT may promote international–interregional–intercultural understanding.

More specifically, Weaver (1993) has analysed the potential benefits of an AT design from the perspective of accommodation, attractions, market, economic impact, and regulation (Table 1.1). This more sensitive approach to tourism development strives to satisfy the needs of local people, tourists, and the resource base in a complementary rather than a competitive manner.

Some researchers, however, are quick to point out that as an option to mass tourism, full-fledged alternative tourism cannot replace conventional tourism simply because of mass tourism's varied and many-sided associated phenomena (Cohen 1987). Instead, it is more realistic to concentrate efforts in attempts to reform the worst prevailing situations, not the

**Table 1.1** *Potential benefits derived from an alternative tourism strategy*

*Accommodation*

- Does not overwhelm the community.
- Benefits (jobs, expenditures) are more evenly distributed.
- Less competition with homes and businesses for the use of infrastructure.
- A larger percentage of revenues accrue to local areas.
- Greater opportunity for local entrepreneurs to participate in the tourism sector.

*Attractions*

- Authenticity and uniqueness of community is promoted and enhanced.
- Attractions are educational and promote self-fulfilment.
- Locals can benefit from existence of the attractions even if tourists are not present.

*Market*

- Tourists do not overwhelm locals in numbers; stress is avoided.
- 'Drought/deluge' cycles are avoided, and equilibrium is fostered.
- A more desirable visitor type.
- Less vulnerability to disruption within a single major market.

*Economic impact*

- Economic diversity is promoted to avoid single-sector dependence.
- Sectors interact and reinforce each other.
- Net revenues are proportionally higher; money circulates within the community.
- More jobs and economic activity are generated.

*Regulation*

- Community makes the critical development/strategy decisions.
- Planning to meet ecological, social, and economic carrying capacities.
- Holistic approach stresses integration and well-being of community interests.
- Long-term approach takes into account the welfare of future generations.
- Integrity of foundation assets is protected.
- Possibility of irreversibilities is reduced.

*Source*: Weaver (1993)

development of alternatives. Butler (1990) feels that mass tourism has not been rejected outright for two main reasons. The first is economic, in that it provides a significant amount of foreign exchange for countries; the second is socio-psychological and relates to the fact that

> many people seem to enjoy being a mass tourist. They actually like not having to make their own travel arrangements, not having to find accommodation when they arrive at a destination, being able to obtain goods and services without learning a foreign language, being able to stay in reasonable, in some cases considerable comfort, being able to eat reasonably familiar food, and not having to spend vast amounts of money or time to achieve these goals.
>
> (Butler 1990: 40)

**Plate 1.2** Alternative tourism can take many forms, including recreational fishing enterprises that maintain local control and small-scale design

## Sustainable development and tourism

The measurement of development (i.e. a nation's stage of socio-economic advancement) has conventionally been accomplished through the implementation of a number of key economic indicators. Among others, these include variables such as protein intake, access to potable water, air quality, fuel, health care, education, employment, GDP and GNP. The so-called 'developed' world (countries like Australia, the USA, Canada, and those of Western Europe) therefore is defined by the existence of these socio-economic conditions, whereby those with more are considered more highly developed (more on development in Chapter 8). Furthermore, one's level of development, either objectively or subjectively, is often equated or synonymous with one's perceived stage of 'civilisation', whereby progress (usually economic) is a key to the relationship between who is civilised and who is not. The *Oxford English Dictionary* defines civilisation as an 'advanced stage of social development', and to civilise as to 'bring out of barbarism, enlighten'. The point to be made is that perhaps our perception of what is developed and what isn't, what is civilised and what isn't, is a matter of debate and one that our more recent approaches to development need to better address. For example, it has been noted that the most developed 20 per cent of the world's population (those in the 'West') are thought to use some 80 per cent of the world's resources with which to achieve development. If it is our goal to have the entire world 'developed' according to this Western paradigm, the planet will be in serious jeopardy. Perhaps in a hundred or two hundred years *Homo sapiens* will look back at Western civilisation as the most barbaric time period in recorded history.

Deming (1996) shares the view that humanity needs to take a good long look at civilisation. She writes that people have an insatiable hunger to see more and more of the planet, and to get closer and closer to its natural attractions. This behaviour surfaces continually in tourism as the tentacles of the tourist seek to push the fine line that exists between acceptable and unacceptable human–wildlife interactions. For example, animal harassment regularly occurs in Point Pelee National Park in Ontario, Canada, as thousands of birders converge on the spring migration of birds in the park. Despite posted warnings, tourists continue to venture off the designated paths in identifying and photographing species. Deming asks: in the face of global warming, diminishing habitat, and massive extinctions, what can it mean to be civilised? Her response is a plea for limits, both social and ecological, in facing the enemy within:

> As Pogo said during the Vietnam War, 'We've seen the enemy and it is us.' Suddenly we are both the invading barbarians and the only ones around to protect the city. Each one of us is at the center of the civilized world and on its edge.
>
> (Deming 1996: 32)

Milgrath (1989) talked of values as fundamental to everything we do (see also Forman 1990). He argues that humans have as a central value their personal desire to preserve their lives. This naturally evolves into a concern and value for other people – a social value. Milgrath suggests that it is inappropriate to elevate the preservation of each human life to a central concern because every person dies and this social preservation can never be realised. He feels instead that we should value the preservation of our ecosystems over society. Beyond the socially oriented values of society, Milgrath says we have given top priority to economic development, the result being that society will not be able to sustain itself over the long term.

Such a focus on instrumental values (something valued as a means to an end), takes us away from the perspective that non-human entities have value in and of themselves, and should exist in their own right. This 'ethic of nature' perspective is one which is more broadly intrinsic and ecocentric (Wearing and Neil 1999).

Sustainable development has been proposed as a model that can have utility in creating the impetus for structural change within society, one that ventures away from a strictly socio-economic focus to one where development 'meets the goals of the present without compromising the ability of future generations to meet their own needs' (World Commission on Environment and Development 1987: 43). As such, the principles of ecology are essential to the process of economic development (Redclift 1987), with the aim of increasing the material standards of people living in the world who are impoverished (Barbier 1987). Even more fundamentally, though, it would seem more inspiring to hope that sustainable development would increase the moral standards of people living everywhere, which might naturally spill over into the realm of economics, which we know is critical to our viability. Tourism's international importance as an engine for economic growth, as well as its potential for growth, makes it particularly relevant to sustainable development. Consequently there is a wealth of literature emerging that is directly related to the sustainability of tourism, however broadly defined.

One of the first action strategies on tourism and sustainability emerged from the Globe '90 conference in British Columbia, Canada. Here, representatives from the tourism industry, government, non-governmental organisations (NGOs), and academia discussed the importance of the environment in sustaining the tourism industry, and how poorly planned tourism developments often erode the very qualities of the natural and human environment that attract visitors. The conference delegates suggested that the goals of sustainable tourism are (1) to develop greater awareness and understanding of the significant contributions that tourism can make to environment and the economy; (2) to promote equity and development; (3) to improve the quality of life of the host community; (4) to provide a high quality of experience for the visitor; and (5) to maintain the quality of the environment on which the foregoing objectives depend. Although their definition of sustainable tourism development was somewhat non-committal (i.e. 'meeting the needs of present tourist and host region while protecting and enhancing opportunity for the future'), a number of good recommendations were developed for policy, government, NGOs, the tourism industry, tourists, and international organisations. For example, the policy section contains 15 recommendations related to how tourism should be promoted, developed, defined, in addition to a series of regional, interregional, and spatial and temporal implications. One of the policy recommendations states that 'sustainable tourism requires the placing of guidelines for levels and types of acceptable growth but does not preclude new facilities and experiences' (Globe '90 1990: 6).

From the perspective of financial prosperity and growth, there is an economic rationale for sustainability; as McCool (1995: 3) asserts, 'once communities lose the character that makes them distinctive and attractive to nonresidents, they have lost their ability to vie for tourist-based income in an increasingly global and competitive marketplace'. In addition, McCool quotes Fallon in suggesting that sustainability is all about the pursuit of goals and measuring progress towards them. No longer is it appropriate to gauge appropriate development by physical output or economic bottom lines; there must also be consideration of social order and justice (see also Hall 1992 and Urry 1992). McCool feels, therefore, that in order for sustainable tourism to be successful, humans must consider the following: (1) how tourists value and use natural environments; (2) how communities are enhanced through tourism; (3) identification of tourism's social and ecological impacts; and (4) management of these impacts.

Accordingly, many researchers and associations have initiated the process of determining and measuring impacts. As outlined above, Globe '90 was one of the initial and integral forces in linking tourism with sustainable development. This was followed by Globe '92 (Hawkes and Williams 1993) and the move from principles to practice in implementing measures of sustainability in tourism. Even so, it was recognised in this conference that there was much work to be done in implementing sustainable principles in tourism, as emphasised by Roy (in Sadler 1992: ix):

> Sustainable tourism is an extension of the new emphasis on sustainable development. Both remain concepts. I have not found a single example of either in India. The closest for tourism is in Bhutan. Very severe control of visitors – 2000 per year – conserves the environment and the country's unique socio-cultural identity. Even there, trekking in the high altitudes, I find the routes littered with the garbage of civilization.

Although many examples exist in the literature on tourism and sustainable development (see Nelson *et al.* 1993), few sustainable tourism projects have withstood the test of time. An initiative that has received some exposure in the literature is the Bali Sustainable Development Project, coordinated through the University of Waterloo, Canada, and Gadjahmada University in Indonesia (see Wall 1993; Mitchell 1994). This is a project that has been applied at a multisectoral level. Tourism, then, is one of many sectors, albeit a prime one, that drives the Balinese economy. Wall (1993) suggests that some of the main conclusions from his work on the project are as follows:

1   Be as culturally sensitive as possible in developing a sustainable development strategy.
2   Work within existing institutional frameworks as opposed to creating new ones.
3   Multi-sectoral planning is critical to a sustainable development strategy and means must be created to allow all affected stakeholders to participate in decision-making.

(See also the work of Cooper (1995) on the offshore islands of the UK and the work of Aylward *et al.* (1996) on the sustainability of the Monteverde Cloud Forest Preserve in

**Table 1.2** *Core indicators of sustainable tourism*

| Indicator | Specific measures |
| --- | --- |
| Site protection | Category of site protection according to IUCN index |
| Stress | Tourist numbers visiting site (per annum/peak month) |
| Use intensity | Intensity of use in peak period (persons/hectare) |
| Social impact | Ratio of tourists to locals (peak period and over time) |
| Development control | Existence of environmental review procedure or formal controls over development of site and use densities |
| Waste management | Percentage of sewage from site receiving treatment (additional indicators may include structural limits of other infrastructural capacity on site, such as water supply) |
| Planning process | Existence of organised regional plan for tourist destination region (including tourism component) |
| Critical ecosystems | Number of rare or endangered species |
| Consumer satisfaction | Level of satisfaction by visitors (survey-based) |
| Local satisfaction | Level of satisfaction by locals (survey-based) |
| Tourism contribution to local economy | Proportion of total economic activity generated by tourism |
| *Composite indices* | |
| Carrying capacity | Composite early warning measure of key factors affecting the ability of the site to support different levels of tourism |
| Site stress | Composite measure of levels of impact on the site (its natural/ cultural attributes due to tourism and other sector cumulative stress) |
| Attractivity | Qualitative measure of those site attributes that make it attractive to tourism and can change over time |

*Source*: Consulting and Audit Canada (1995)

Costa Rica as good examples of tourism and sustainability.) The integration of tourism with other land uses in a region has also been addressed by Butler (1993: 221), who sees integration as 'the incorporation of an activity into an area on a basis acceptable to other activities and the environment within the general goal of sustainable or long-term development'. Butler identified complementarity, compatibility, and competitiveness as variables that could be used as a first step in prioritising land uses, where complementarity leads to a higher degree of integration and competitiveness leads to segregation of the activity relative to other land uses.

Other models have been more unisectoral in their approach to the place of tourism within a destination region. These have tended to identify a range of indicators that identify a sustainable approach or unsustainable approach to the delivery of tourism. Examples include Canova's (1994) illustration of how tourists can be responsible towards the environment and local populations; Forsyth's (1995) overview of sustainable tourism and self-regulation; Moscardo *et al*.'s (1996) look at ecologically sustainable forms of tourism accommodation; and Consulting and Audit Canada's (1995) guide to the development of core and site-specific sustainable tourism indicators (see also Manning 1996). Table 1.2 identifies the core indicators identified in this document. These core indicators (e.g. site protection, stress, use intensity, waste management, and so on) must, according to the report, be used in concert with specific site or destination indicators. This report identifies two categories of this latter group of indicators: (1) supplementary ecosystem-specific indicators (applied to specific biophysical land and water regions), and (2) site-specific indicators, which are developed for a particular site. Table 1.3 provides an overview of some of these 'secondary' ecosystem indicators.

**Table 1.3** *Ecosystem-specific indicators*

| Ecosystem | Sample indicators[a] |
|---|---|
| Coastal zones | Degradation (percentage of beach degraded, eroded)<br>Use intensity (persons per metre of accessible beach)<br>Water quality (faecal coliform and heavy metals counts) |
| Mountain regions | Erosion (percentage of surface area eroded)<br>Biodiversity (key species counts)<br>Access to key sites (hours' wait) |
| Managed wildlife parks | Species health (reproductive success, species diversity)<br>Use intensity (ratio of visitors to game)<br>Encroachment (percentage of park affected by unauthorised activity) |
| Ecologically unique sites | Ecosystem degradation (number and mix of species, percentage area with change in cover)<br>Stress on site (number of operators using site)<br>Number of tourist sitings of key species (percentage success) |
| Urban environments | Safety (crime numbers)<br>Waste counts (amounts of rubbish, costs)<br>Pollution (air pollution counts) |
| Cultural sites (built) | Site degradation (restoration/repair costs)<br>Structure degradation (precipitation acidity, air pollution counts)<br>Safety (crime levels) |
| Cultural sites (traditional) | Potential social stress (ratio average income of tourists/locals)<br>In season sites (percentage of vendors open year round)<br>Antagonism (reported incidents between locals and tourists) |
| Small islands | Currency leakage (percentage of loss from total tourism revenues)<br>Ownership (percentage foreign ownership of tourism establishments)<br>Water availability (costs, remaining supply) |

*Source*: Manning (1996)
*Note*:
[a] These ecosystem-specific indicators are merely suggested, and act as supplements to core indicators

Some publications have discussed tourism and sustainability from the perspective of codes of ethics (codes of ethics are discussed at length in Chapter 9). While indicators are variables that are identified and used to measure and monitor tourism impacts, codes of ethics or conduct are lists designed to elicit a change in behaviour of particular stakeholder groups; a form of compliance for acceptable behaviour at a tourism setting. The *Beyond the Green Horizon* paper on sustainable tourism (Tourism Concern 1992) is a good example of this form of education. To Tourism Concern, sustainable tourism is:

> tourism and associated infrastructures that, both now and in the future: operate within natural capacities for the regeneration and future productivity of natural resources; recognise the contribution that people and communities, customs and lifestyles, make to the tourism experience; accept that these people must have an equitable share in the economic benefits of tourism; are guided by the wishes of local people and communities in the host areas.

①  **Using Resources Sustainably**
The conservation and sustainable use of resources – natural, social and
cultural – is crucial and makes long-term business sense.

②  **Reducing Over-Consumption and Waste**
Reduction of over-consumption and waste avoids the costs of restoring
long-term environmental damage and contributes to the quality of tourism.

③  **Maintaining Diversity**
Maintaining and promoting natural, social and cultural diversity is essential
for long-term sustainable tourism, and creates a resilient base for the industry.

④  **Integrating Tourism into Planning**
Tourism development which is integrated into a national and local
strategic planning framework, and which undertakes EIAs, increases the
long-term viability of tourism.

⑤  **Supporting Local Economies**
Tourism that supports a wide range of local economic activities, and which
takes environmental costs/values into account, both protects those economies
and avoids environmental damage.

⑥  **Involving Local Communities**
The full involvement of local communities in the tourism sector not only
benefits them and the environment in general but also improves the quality
of the tourism experience.

⑦  **Consulting Stakeholders and the Public**
Consultation between the tourism industry and local communities,
organisations and institutions is essential if they are to work alongside each other
and resolve potential conflicts of interest.

⑧  **Training Staff**
Staff training which integrates sustainable tourism into work practices,
along with recruitment of local personnel at all levels, improves the quality
of the tourism product.

⑨  **Marketing Tourism Responsibly**
Marketing that provides tourists with full and responsible information
increases respect for the natural, social and cultural environments of
destination areas and enhances customer satisfaction.

⑩  **Undertaking Research**
Ongoing research and monitoring by the industry using effective data
collection and analysis is essential in solving problems and bringing benefits
to destinations, the industry and consumers.

**Figure 1.1** Principles for sustainable tourism

*Source*: Tourism Concern (1992)

Nothing is measured but 'rules' are stated for the purpose of prompting or reinforcing this appropriate behaviour. A further elaboration of Tourism Concern's ten guiding principles can be found in Figure 1.1.

The Tourism Industry Association of Canada (1995) joined forces with the National Round Table on the Environment and the Economy in creating a document that demonstrates commitment and responsibility to protecting the environment through cooperation with other sectors and governments at all levels. Their sustainable tourism guidelines were developed for tourists, the tourism industry, industry associations, accommodation, food services, tour operators, and Ministries of Tourism. Each of these sections contains appropriate guidelines that deal with policy and planning; the tourism experience; the host community; development; natural, cultural, and historic resources; conservation of natural resources; environmental protection; marketing; research and education; public awareness; industry cooperation; and the global village.

A final publication that merits attention in this section is the work of the Federation of Nature and National Parks of Europe (1993). Its comprehensive look at sustainable tourism in Europe's nature and national parks provides good insight into the challenge of implementing sustainability in that part of the world. Many of the protected areas in Europe are situated in rural working landscapes (e.g. England, Wales, Luxembourg) and must contend with different pressures as compared with some of the larger and less densely populated areas surrounding the protected areas of Australia and New Zealand, Canada, and the United States. However, Europe also contains many large national parks and biosphere reserves that are maintained accordingly. In both cases (rural and wilderness environments) policy-makers and practitioners are charged with the task of implementing sustainable tourism in these varied settings. The European national parks document recognises that people must be able to improve the quality of their lives, maintain jobs, improve their economy, enjoy their cultures, and promote harmony between cultures. These must be accomplished with an eye to environmental education, political support for the environment, and the protection of heritage values through restorative projects and direct practical help.

Sustainable tourism, however, is not without its critics. Hunter (1995), for example, suggests that the current approach to sustainable tourism development is one that is flawed because it condones the planning and management of tourism in a manner inconsistent with the design of sustainable development. In particular, tourism does not adequately address issues of geographical scale and intersectoral cooperation which are so important to achieving sustainable development. Furthermore, Macbeth (1994) calls attention to the fact that sustainable tourism is more reactionary than proactive in nature. Macbeth suggests that 'the history of capitalism is full of examples of how reactionary tendencies are easily co-opted by capitalism to sustain its own existence, thus extending the status quo of exploitive relations rather than overthrowing them' (ibid.: 44). This will continue to occur, according to Macbeth, unless the present form of capitalism is overcome.

McKercher (1993a) feels that tourism is vulnerable to losing sustainability for four main reasons. First, tourism is not recognised as a natural resource-dependent industry; second, the tourism industry is invisible, especially in urban areas; third, tourism is electorally weak, with little support in government; and fourth, there is a distinct lack of leadership driving the industry, which ultimately makes tourism vulnerable to attacks from other land users. McKercher cites the example of resource use in northern Ontario as a case in point. In this region the economy has been dominated politically by the large extractive industries (forestry and mining). The disaggregated structure of the tourism industry in Ontario's north (predominantly outfitters and lodges) prevents it from having any political decision-making influence at all.

Other critical reviews of tourism and sustainability include Goodall and Cater's (1996) belief that sustainable tourism will probably not be achieved, despite the most committed environmental performance, and Durr's (1995) work illustrating that sustainable tourism development is unlikely to occur unless the people of rural tourism communities work together to make it happen. There appears to be certain agreement that if sustainability is to occur at all, it must be done at the local level, and perhaps shaped loosely by a broader national or international policy. The notions of policy and local participation have been examined by Laarman and Gregersen (1994), who feel that sustainable nature tourism policy

## CASE STUDY 1.1

### *Sustainable tourism and the Green Villages of Austria*

Understanding the vital link between landscape and tourism, Austria has embarked upon a policy of sustainable tourism with the aim of preservation and an overall improvement in the quality of the natural environment. Specifically, the following measures have been proposed:

1  A straightening out of the demand curve to avoid peak demands and burdens;
2  reducing the consumption of space for tourism;
3  preservation of natural landscapes;
4  cooperation with other industries, in particular agriculture and forestry;
5  professionalism within the industry; and
6  a changing of the behaviour of tourists.

One of the most significant programmes in Austria is the Green Village endeavour, which is designed to allow communities to accommodate the growing demands of tourism in a sustainable way. Towns are encouraged to incorporate solar panels in their heating, restrict building height to no more than three storeys, keep parking places a minimum of 80 metres away from buildings to eliminate noise and fumes, keep motorways at least 3 km away from Green Villages, restrict vehicular traffic through villages, designate cycle paths, recycle, restrict building to the town site only, eliminate single-crop farming in adjacent farmlands, discriminate in favour of sustainable craftsmen, build hotels using natural products, insist that farmers be able to sell their products locally, and use local, natural pharmaceuticals. Such a philosophy, it is thought, will benefit both communities and the tourism industry.

must include the following three areas: (1) national support and advanced planning; (2) appropriate pricing and revenue policies; and (3) local participation and benefits.

Some authors have identified unique sustainable tourism market segments, suggesting that each uses the natural environment, has long-term economic benefits, involves continuous environmental protection, and stimulates local community development. Although each of the foregoing variables can be used to define sustainable tourism, it is important to realise that AT may be no more sustainable than Cancún and the nearby Cozumel Island if improperly managed, and there are decidedly few examples proven to be properly managed. It is therefore potentially dangerous to look at sustainable tourism as a specific market, instead of from site-specific or regional case studies.

Figure 1.2 illustrates that sustainability has to be more than simply one aspect of the industry (e.g. accommodation) working in a sustainable way. The illustration recognises that in essence the tourism industry experiences a tremendous degree of fragmentation by virtue of the fact that consistency in sustainability is not likely to be found across all sectors. The aim, then, for sustainability is to ensure that all aspects of the industry are working in concert. In addition, the figure incorporates the notion of both human and physical elements working within each of the four sectors; that is, the fact that the people working at a physical attraction very much dictate the extent to which sustainability is achieved at the site.

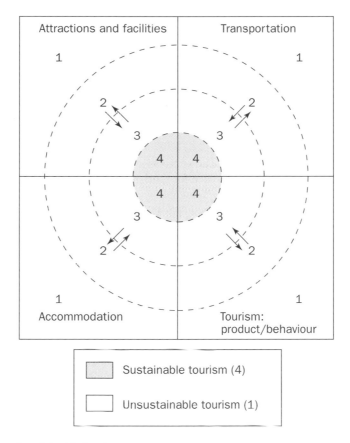

Figure 1.2  Degrees of sustainable tourism

## Conceptualising tourism and sustainability

In the previous discussion, mass tourism, AT, and sustainable tourism have been analysed individually. The relationship that they share, however, can be more fully appreciated in the conceptual framework shown in Figure 1.3. In a general sense, the illustration provides a good sense of the relative size of mass tourism and alternative tourism according to the corresponding circles in the diagram. Although mass tourism may be said to be predominantly unsustainable, more recently new and existing developments in the industry have attempted to encourage more sustainable practice through various measures, some of which include the controlled use of electricity, a rotating laundry schedule, and the disposal of wastes (the arrow indicates that there is a move towards an increasing degree of sustainability in this sector). On the other hand, the illustration indicates that most forms of alternative tourism are sustainable in nature (in theory). The AT sphere is shown to comprise two types of tourism, socio-cultural tourism and ecotourism. Socio-cultural AT includes, for example, rural or farm tourism, where a large portion of the touristic experience is founded upon the cultural milieu that corresponds to the environment in which farms operate. Ecotourism, however, involves a type of tourism that is less socio-cultural in its orientation, and more dependent upon nature and natural resources as the primary component or motivator of the trip, hence the division within the AT sphere (see Chapter 2 for an extensive overview of

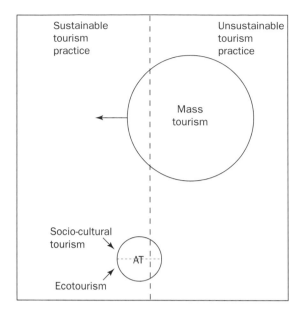

**Figure 1.3** Tourism relationships

*Source*: Adapted from Butler (1996) in Weaver (1998)

ecotourism definitions and research). The focus of this book is on this lower half of the AT circle only, and the belief that ecotourism is distinct from mass tourism and various other forms of AT.

## Conclusion

Given the emergent wealth of literature on tourism, AT and sustainability, one cannot help but reflect on where we have been, but also on what is missing. Researchers have written countless reports, codes of conduct, guidelines, and indicators, but few attempts and successes can be drawn upon to point the direction, as illustrated in Butler's (1991) sobering analysis of tourism, environment, and sustainable development. Perhaps it is still too early, and these anticipated successes will be the next stage of tourism's rather tenuous relationship with sustainability. However, there is a pressing need to move beyond the rhetoric; beyond the table and into practice, for according to Leslie (1994: 35) 'tourism in terms of practice in developing countries is a mirror in which we see the possible mistakes identifiable retrospectively in the development and wasteful consumerist lifestyles of the west'. One cannot discount the importance of the public in operationalising AT and sustainable tourism practices. People have the power to demand goods and services that are developed and presented in an ecologically friendly manner. The public must demand accountability of tourism products, and tourism service providers must demonstrate an adherence to an appropriate vision in striving for meritorious achievements in the area of sustainable development. In doing so, it is those who achieve such lofty levels who may ultimately prosper financially.

## Summary questions

1    Why is tourism so hard to define?
2    What are some of the characteristics that have been used to define attractions?
3    What are the basic differences between mass tourism and alternative tourism?
4    Why is a discussion of values so important in the understanding of sustainable development and sustainable tourism?
5    How do core indicators of sustainable tourism differ from ecosystem-specific indicators?

# **2** Ecotourism and ecotourists

As an extension or outgrowth of alternative tourism (AT), ecotourism has grown as a consequence of the dissatisfaction with conventional forms of tourism which have, in a general sense, ignored social and ecological elements of foreign regions in favour of a more anthropocentric and strictly profit-centred approach to the delivery of tourism products. This chapter presents an in-depth analysis of ecotourism from the perspective of history, definition, and linkages to other related forms of tourism (i.e. adventure tourism and cultural tourism). A brief discussion of tourism typologies is offered as a means by which to understand better the documented differences that exist between ecotourists and other types of travellers.

## Ecotourism

Until recently, there has been some confusion surrounding the etymology or origin of the term 'ecotourism', as evident in the tremendous volume of literature on the topic. For example, Orams (1995) and Hvenegaard (1994) write that the term can be traced back only to the late 1980s, while others (Higgins 1996) suggest that it can be traced to the late 1970s through the work of Miller (see Miller 1989) on ecodevelopment. One of the consistent themes emergent in the literature supports the fact that Ceballos-Lascuráin was the first to coin the phrase in the early 1980s (see Thompson 1995). He defined it as 'traveling to relatively undisturbed or uncontaminated natural areas with the specific objective of studying, admiring, and enjoying the scenery and its wild plants and animals, as well as any existing cultural manifestations (both past and present) found in these areas' (Boo 1990: xiv). In an interview (van der Merwe 1996), Ceballos-Lascuráin illustrates that his initial reference to the phrase occurred in 1983, while he was in the process of developing PRONATURA, an NGO in Mexico. Recently, however, the term has been traced further back to the work of Hetzer (1965), who used it to explain the intricate relationship between tourists and the environments and cultures in which they interact. Hetzer identified four fundamental pillars that needed to be followed for a more responsible form of tourism. These included: (1) minimum environmental impact; (2) minimum impact on – and maximum respect for – host cultures; (3) maximum economic benefits to the host country's grassroots; and (4) maximum 'recreational' satisfaction to participating tourists. The development of the concept of ecotourism grew, according to Hetzer (personal communication, October 1997), as a culmination of dissatisfaction with governments' and society's negative approach to development, especially from an ecological point of view. Nelson (1994) also adopts this particular stand in illustrating that the idea of ecotourism is in fact an old one, which manifested itself during the late 1960s and early 1970s when researchers became concerned over inappropriate use of natural resources. Nelson suggests that the term 'eco-development' was introduced as a means by which to reduce such development.

In other related research, Fennell (1998) found evidence of Canadian government 'ecotours' which were operational during the mid-1970s. These ecotours centred around the Trans-Canada Highway and were developed on the basis of different ecological zones found along the course of the highway. The first of these encounters was developed in 1976. Based on the ecozone concept, they were felt to be rather progressive for the time despite the lack of a focused look at low impact, sustainability, community development, and the moral philosophy labels that are attached to ecotourism in the 1990s. The ecotours were developed at a time when the Canadian government felt it important to allow Canadian and foreign travellers to appreciate the human–land relationship in Canada, through the interpretation of the natural environment. Although a set definition of ecotourism was not provided, each of the ecotour guides contains the following foreword:

> Ecotours are prepared by the Canadian Forestry Service to help you, as a traveller, understand the features of the landscape you see as you cross the country. Both natural and human history are described and interpreted. The route covered by the Ecotours is divided into major landscape types, or Ecozones, and a map of each Ecozone shows the location of interesting features (identified by code numbers). While most features can be seen from your car, stops are suggested for some of them. Distances between points of interest are given in kilometres. Where side trips are described, distances are given to the turnoff from the highway. You will derive the maximum value from this Ecotour if you keep a record of the distance travelled and read the information on each point of interest before reaching it.
>
> (Fennell 1998: 232)

Fennell goes on to suggest that ecotourism most likely has a convergent evolution, 'where many places and people independently responded to the need for more nature travel opportunities in line with society's efforts to become more ecologically minded' (Fennell 1998: 234), as also suggested by Nelson (see above). This evidence comes at a time when researchers have been struggling to find common ground between ecotourism and its relationship to other forms of tourism, related and unrelated. (For other early references on ecotourism see Mathieson and Wall 1982; Romeril 1985.)

There seems to be universal acceptance of the fact that ecotourism was viable long before the 1980s in practice, if not in name. For example, Blangy and Nielson (1993) illustrate that the travel department of the American Museum of Natural History has conducted natural history tours since 1953. Probably the finest examples of the evolution of ecotourism can be found in the African wildlife-based examples of tourism developed in the early twentieth century and, to some, the nature tourism enterprises of the mid-nineteenth century (Wilson 1992). There are examples in the literature that illustrate that human beings, at least since the Romantic period, have travelled to the wilderness for the intrinsic nature of the experience. Nash writes that during the nineteenth century many people travelled both in Europe and North America for the primary purpose of enjoying the outdoors, as illustrated in the following passage:

> Alexis de Tocqueville resolved to see wilderness during his 1831 trip to the United States, and in Michigan Territory in July the young Frenchman found himself at last on the fringe of civilization. But when he informed the frontiersmen of his desire to travel for pleasure into the primitive forest, they thought him mad. The Americans required considerable persuasion from Tocqueville to convince them that his interests lay in matters other than lumbering or land speculation.
>
> (Nash 1982: 23)

Tocqueville was after something that we consider as an essential psychological factor in travel: novelty. Nash (1982) credits the intellectual revolution in the eighteenth and nineteenth centuries as the push needed to inspire the belief that unmodified nature could act as a deep spiritual and psychological tonic. It required the emergence of a group of affluent and cultured persons who largely resided in urban environments to garnish this appreciation (e.g. Jean-Jacques Rousseau and John Ruskin). For these people, Nash (ibid.: 347) writes, 'wilderness could become an intriguing novelty and even a deep spiritual and psychological need'. In the United States, the sentiment at the time was not as strong as it was in Europe, where 'as late as the 1870s almost all nature tourists on the American frontier continued to be foreigners' (ibid.: 348).

When Americans did start travelling to the wild parts of their country it was the privileged classes that held the exclusive rights. A trip to Yellowstone in the 1880s, according to O'Gara (1996), was about three times as expensive as travel to Europe at the time. There was no question that those from the city were especially taken with Yellowstone's majesty, but their mannerisms left much to be desired, as is evident in an account of such tourists by Rudyard Kipling (ibid.: 56):

> It is not the ghastly vulgarity, the oozing, rampant Bessemer steel self-sufficiency and ignorance of the men that revolts me, so much as the display of these same qualities in the womenfolk. . . . All the young ladies . . . remarked that [Old Faithful] was 'elegant' and betook themselves to writing their names in the bottoms of the shallow pools. Nature fixes the insult indelibly, and the after-years will learn that 'Hattie,' 'Sadie,' 'Mamie,' 'Sophie,' and so forth, have taken out their hairpins and scrawled in the face of Old Faithful.

## Concepts and variables: definitions of ecotourism

Given the ambiguity associated with the historical origins of ecotourism, the purpose of the present section is to identify the key features, concepts, and principles of the term, especially the link between nature tourism (or nature-oriented tourism) and ecotourism. For example, Laarman and Durst, in their early reference to ecotourism, defined it as a nature tourism in which the 'traveler is drawn to a destination because of his or her interest in one or more features of that destination's natural history. The visit combines education, recreation, and often adventure' (Laarman and Durst 1987: 5). In addition, these authors were perhaps the first to make reference to nature tourism's hard and soft dimensions, based on the physical rigour of the experience and also the level of interest in natural history (Figure 2.1). Laarman and Durst suggested that scientists would in most likelihood be more dedicated than casual in their pursuit of ecotourism, and that some types of ecotourists would be more willing to endure hardships than others in order to secure their experiences. The letter 'B' in Figure 2.1 identifies a harder ecotourism experience based on a more difficult or rigorous experience, and also based on the dedication shown by the ecotourist relative to the interest in the activity. (See Weaver [2001a] and Fennell [2002a] for a more in-depth discussion of the hard and soft dimensions of ecotourism.)

A subsequent definition by Laarman and Durst (1993) identifies a conceptual difference between ecotourism and nature tourism. In recognising the difficulties in defining nature tourism, they establish both a narrow and broad scope to its definition. Narrowly, they say, it refers to operators running nature-oriented tours; however, broadly it applies to tourism's use of natural resources including beaches and country landscapes. In their research they define nature tourism as 'tourism focused principally on natural resources such as relatively

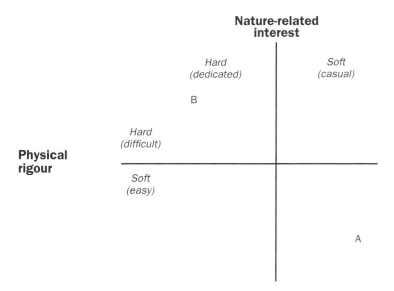

**Figure 2.1** Hard and soft dimensions of ecotourism

*Source*: Laarman and Durst (1987)

undisturbed parks and natural areas, wetlands, wildlife reserves, and other areas of protected flora, fauna, and habitats' (ibid.: 2). Given this perspective, there appears to be some consensus mounting in the literature that describes ecotourism as one part of a broader nature-based tourism. This becomes evident in the discussion by Goodwin (1996: 287), who wrote that *nature tourism*

> encompasses all forms of tourism – mass tourism, adventure tourism, low-impact tourism, ecotourism – which use natural resources in a wild or undeveloped form – including species, habitat, landscape, scenery and salt and fresh-water features. Nature tourism is travel for the purpose of enjoying undeveloped natural areas or wildlife.

And conversely, that *ecotourism* is

> low impact nature tourism which contributes to the maintenance of species and habitats either directly through a contribution to conservation and/or indirectly by providing revenue to the local community sufficient for local people to value, and therefore protect, their wildlife heritage area as a source of income.
>
> (Goodwin 1996: 288)

Apart from the differences apparent in the work of Goodwin, some of the key variables or principles that separate ecotourism from its more broad-based nature counterpart include an educative component and a sustainability component (Blamey 1995), and the ethical nature of the experience (Kutay 1989; Wight 1993a; Hawkes and Williams 1993; Wallace and Pierce 1996).

More recently the Quebec Declaration (UNEP/WTO 2002), at the penultimate meeting of the International Year of Ecotourism (in 2002), suggested that five distinct criteria be used to define ecotourism, namely: nature-based product, minimal impact management, environmental education, contribution to conservation and contribution to community.

## CASE STUDY 2.1

### *Nature tourism in Texas*

Ecotourism in Texas provides an excellent example of how ecotourism is subsumed by nature tourism. In Texas, nature tourism is defined as 'discretionary travel to natural areas that conserves the environmental, social and cultural values while generating an economic benefit to the local community' (Texas Parks and Wildlife 1996: 2). Although hunting and fishing are reported to be traditional mainstays of nature tourism in Texas, the Task Force on Nature Tourism states that non-consumptive activities such as bird and wildlife watching, nature study and photography, biking, camping, rafting, and hiking have experienced the greatest growth over the past few years. In Texas, the task force reports that tourism is the third largest industry in Texas, generating $23 billion annually, with the potential to replace oil and gas, and manufacturing as the highest income earner by the turn of the century.

The Great Texas Coastal Birding Trail is a key ecotourism attraction in the Lone Star State. The goal of this trail is to increase the opportunities for nature tourism in the coastal communities of Texas, in addition to conveying the value of conservation to people living in the region. Although the trail was conceived by the Texas Parks and Wildlife Department, it was made possible by transportation enhancement funds and the Texas Department of Transportation. The trail utilises existing transportation infrastructure (viewing platforms and boardwalks are being added on a continual basis) in creating recreational, economic, and educational opportunities for local people and tourists alike. Upon completion in 1998, it comprised three main sections and span over 600 miles of coastline, incorporating some 300 birding stops in nine wildlife refuges, eleven state parks, one national seashore, and several city and county preserves.

In view of the relationship between the two concepts, there does seem to be some merit in linking ecotourism to nature tourism, given the tremendous variety of nature-related tourism interests. However, there is also ambiguity in separating nature tourism from other forms of tourism, all of which rely upon the use of natural resources. Even mass, resort-based tourism relies upon undeveloped resources (i.e. beaches and the ocean) as a central component of the product and experience (see Weaver 2001a). An important aspect of Goodwin's discussion includes the fact that not all types of nature tourism are necessarily compatible with each other or the environment. Examples include hunting and birdwatching, which place different demands on the resource base, and users, and which have a different managerial focus (if managed at all) based on the consumptive nature of these activities. The importance of management in guiding the ecotourism product was central to the work of Fennell and Eagles (1990), who included the resource tour as the principal component of the ecotourism experience; the service industry, including tour operation, government policy, resource management and community development; and the visitor experience, based on marketing, visitor management and visitor attitudes (Figure 2.2). In their work, Fennell and Eagles recognised the value in management from the visitors' perspective, but also from the community and resource-based perspectives (see also Hvenegaard 1994 for a good description of ecotourism conceptual frameworks).

The variability of approaches used to define ecotourism has quite naturally given way to many comprehensive approaches used to understand the term. In her approach to the concept, Ziffer (1989) acknowledges the variety of descriptive terms such as 'nature travel', 'adventure travel' and 'cultural travel' which are largely activity based; and also the

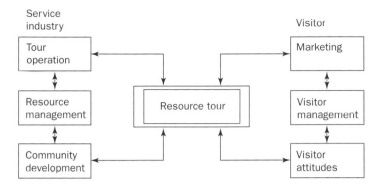

**Figure 2.2** Ecotourism framework

*Source*: Fennell and Eagles (1990)

value-laden terms such as 'responsible', 'alternative' and 'ethical' tourism, which underscore the need to consider impacts and the consequences of travel. Ziffer feels that nature tourism, while not necessarily ecologically sound in principle, concentrates more on the motivation and the behaviour of the individual tourist. Conversely, ecotourism is much more difficult to attain owing to its overall comprehensiveness (the need for planning and the achievement of societal goals). She defines ecotourism as follows:

> a form of tourism inspired primarily by the natural history of an area, including its indigenous cultures. The ecotourist visits relatively undeveloped areas in the spirit of appreciation, participation and sensitivity. The ecotourist practises a non-consumptive use of wildlife and natural resources and contributes to the visited area through labor or financial means aimed at directly benefiting the conservation of the site and the economic well-being of the local residents. The visit should strengthen the ecotourist's appreciation and dedication to conservation issues in general, and to the specific needs of the locale. Ecotourism also implies a managed approach by the host country or region which commits itself to establishing and maintaining the sites with the participation of local residents, marketing them appropriately, enforcing regulations, and using the proceeds of the enterprise to fund the area's land management as well as community development.
> (Ziffer 1989: 6)

Like Ziffer's, the following definition by Wallace and Pierce (1996: 848) is also quite comprehensive, acknowledging the importance of a broad number of variables. To these authors, ecotourism is

> travel to relatively undisturbed natural areas for study, enjoyment, or volunteer assistance. It is travel that concerns itself with the flora, fauna, geology, and ecosystems of an area, as well as the people (caretakers) who live nearby, their needs, their culture, and their relationship to the land it [*sic*] views natural areas both as 'home to all of us' in a global sense ('eco' meaning home) but 'home to nearby residents' specifically. It is envisioned as a tool for both conservation and sustainable development – especially in areas where local people are asked to forgo the consumptive use of resources for others.

Wallace and Pierce suggest that this tourism may be said to be true ecotourism if it addresses six principles:

1    It entails a type of use that minimises negative impacts to the environment and to local people.
2    It increases the awareness and understanding of an area's natural and cultural systems and the subsequent involvement of visitors in issues affecting those systems.
3    It contributes to the conservation and management of legally protected and other natural areas.
4    It maximises the early and long-term participation of local people in the decision-making process that determines the kind and amount of tourism that should occur.
5    It directs economic and other benefits to local people that complement rather than overwhelm or replace traditional practices (farming, fishing, social systems, etc.).
6    It provides special opportunities for local people and nature tourism employees to utilise and visit natural areas and learn more about the wonders that other visitors come to see.

Both Ziffer and Wallace and Pierce recognise that for ecotourism to exist it must strive to reach very lofty goals. By comparison, the Ecotourism Society advocates a much more general definition of the term, one that advocates a 'middle-of-the-road' or passive position (see Orams 1995), and one which is more easily articulated. This organisation defines ecotourism as 'responsible travel to natural areas which conserves the environment and improves the welfare of local people' (Western 1993: 8). Preece *et al.* (1995) used the Australian National Ecotourism Strategy definition of ecotourism in their overview of biodiversity and ecotourism, which is also one that is quite general in nature. The strategy defines ecotourism as nature-based tourism that involves education and interpretation of the natural environment and is managed to be ecologically sustainable.

Although quite straightforward, these definitions leave much to the interpretation of the reader. As identified in Table 2.1, definitions of ecotourism over the years have emphasised different aspects of nature, relationships with local people, conservation and preservation, and so on. The main issue confronting researchers who are bent on structuring definitions is either to include a whole series of principles and variables related to the term, or to try to isolate specific variables which can be used to best represent the overall concept. The fact of the matter is that, either implicitly or explicitly, such variables or principles need to be more effectively used to observe, measure, and evaluate what is and what is not ecotourism.

In line with the previous discussion, Bottrill and Pearce (1995) point to the fact that the rich array of definitions of ecotourism have done little to clarify its meaning. Instead they opt for a means by which to operationalise ecotourism through the development of a set of measurable key elements for tourists, service providers and land managers (see also Wallace and Pierce 1996, and their evaluation of ecotourism principles in Amazonas). In an analysis of 22 ecotourism ventures only five were classified as ecotourism using the following criteria: motivation (physical activity, education, participation), sensitive management, and protected area status. The authors submit that more work should follow to further define and modify the points and criteria raised. Their paper quite nicely addressed the need to move beyond definition to a position where ecotourism operators should be open to ethical and operational scrutiny by the public and other concerned stakeholders. This leads to the contentions of Miller and Kaye (1993: 37), who suggest that 'the merits or deficiencies of ecotourism . . . are not to be found in any label *per se*, but in the quality and intensity of specific environmental and social impacts of human activity in an ecological system'.

Some researchers have discussed the use of time as a mechanism by which to classify ecotourists. For example, Ballantine and Eagles (1994) suggested that ecotourists could be defined on the basis of an intention to learn about nature, an intention to visit undisturbed areas, and a commitment of at least 33 per cent of their time to the first two criteria. Blamey (1995) sees some inherent problems with the implementation of such a measure. The time factor advocated by Ballantine and Eagles may be applicable in the safari ecotourism settings

**Table 2.1** *Comparisons of selected ecotourism and nature tourism definitions*

| Main principles of definition[a] | Definitions | | | | | | | | | | | | | | |
|---|---|---|---|---|---|---|---|---|---|---|---|---|---|---|---|
| | 1 | 2 | 3 | 4 | 5 | 6 | 7 | 8 | 9 | 10 | 11 | 12 | 13 | 14 | 15 |
| Interest in nature | ✔ | ✔ | | ✔ | ✔ | ✔ | ✔ | | ✔ | ✔ | | | | ✔ | ✔ |
| Contributes to conservation | | ✔ | | ✔ | ✔ | ✔ | ✔ | ✔ | ✔ | | | | ✔ | ✔ | ✔ |
| Reliance on parks and protected areas | ✔ | | ✔ | | ✔ | ✔ | | ✔ | ✔ | | | | ✔ | ✔ | ✔ |
| Benefits local people/long-term benefits | | ✔ | ✔ | | ✔ | ✔ | ✔ | | ✔ | | | | ✔ | ✔ | ✔ |
| Education and study | ✔ | ✔ | ✔ | | | ✔ | | | | ✔ | | | | ✔ | ✔ |
| Low impact/non-consumptive | | | | | ✔ | | | | | | | ✔ | ✔ | ✔ | ✔ |
| Ethics/responsibility | | | | ✔ | | | | | ✔ | ✔ | | | | | ✔ |
| Management | | | | | ✔ | | | ✔ | | | ✔ | | | | ✔ |
| Sustainable | | | | | | | | ✔ | | | ✔ | | | ✔ | ✔ |
| Enjoyment/appreciation | ✔ | | | | | ✔ | | | | | | | | ✔ | |
| Culture | ✔ | | | | | ✔ | | | | | | | | ✔ | |
| Adventure | | | ✔ | | | | | | | | | | | | |
| Small scale | | | | | | | | | | | | ✔ | | | ✔ |

*Sources*: 1 Ceballos-Lascuráin (1987); 2 Laarman and Durst (1987)[b]; 3 Halbertsma (1988)[b]; 4 Kutay (1989); 5 Ziffer (1989); 6 Fennell and Eagles (1990); 7 CEAC (1992); 8 Valentine (1993); 9 The Ecotourism Society; 10 Western; 11 Australian National Ecotourism Strategy; 12 Brandon (1996); 13 Goodwin (1996); 14 Wallace and Pierce (1996); 15 The present study

*Notes*:
[a] Variables ranked by frequency of response
[b] Nature tourism definitions

of Africa, but could be more problematic in less-structured ecotourism settings and situations. For instance, Blamey asks if a ten-minute guided nature tour qualifies as a day ecotourism visit. Yet one can't help but grapple for acceptable definitions of ecotourism, as Ballantine and Eagles and others have done; the same has occurred with the meaning of 'tourism' in the past, and similarly for many other social science concepts. While the tourism literature has seen fit to define 'tourism' under many different circumstances – time, space, economics, whole systems models – the same will likely occur for ecotourism. It depends on who is operationalising the concept, and for what purpose. (See also Buckley 1994.)

Given the foregoing discussion, I have elected to provide my own definition of ecotourism. In light of the review of a significant number of definitions, it is based on what are felt to be the most important aspects of the phenomenon and on the need to be concise in including such principles. It is as follows:

> *Ecotourism is a sustainable form of natural resource-based tourism that focuses primarily on experiencing and learning about nature, and which is ethically managed to be low-impact, non-consumptive, and locally oriented (control, benefits, and scale). It typically occurs in natural areas, and should contribute to the conservation or preservation of such areas.*

Admittedly, this is more comprehensive than other definitions, but it strives to recognise that having been identified as a separate form of tourism, ecotourism must be classified and defined as such in order to maintain an element of homogeneity. This feeling is not expressed without some hesitation based on the writing of some other authors who imply that most people who demand ecotourism want a softer, easily accessible, front-country type of experience (Kearsley 1997 in Weaver 1998) and that the 'popular' form of ecotourism demands mechanised transport, easy accessibility, and a high level of services (Queensland Draft Ecotourism Strategy in Weaver 1998). The relationship between this very soft form of ecotourism and other types of tourism has not been fully examined. The belief in this book is that much more than simply demand must go into the understanding of ecotourism and ecotourists. In an attempt to stay clear of this fine line, a harder stance on ecotourism is adopted. Furthermore, a stricter definition of ecotourism begs for the employment of measurable indicators in determining what is and is not ecotourism. (See Orams 1995 for a description of the hard–soft path ecotourism continuum.) An early example of this type of thinking can be seen through the efforts of Shores (1992), who identified the need for higher standards in the ecotourism industry through the implementation of a scale to measure the level of achievement according to the principles of ecotourism. The scale ranges from 0 (travellers made aware of the fragility of the environment in a general capacity) to 5 (a trip where the entire system was operating in an environmental way).

The reader will most certainly recognise the absence of culture as a fundamental principle of ecotourism in the aforementioned definition. This definition views culture only inasmuch as the benefits from ecotourism accrue to local people, recognising that culture, whether exotic or not, is part of any tourism experience. If culture was a primary theme of ecotourism then it would be cultural tourism. There is no doubt that culture can be part of the ecotourism experience; the point is, however, that it is more likely to be a secondary motivation to the overall experience, not primary as in the case of nature and natural resources. For example, in a study completed by Fennell (1990) it was found that there was no statistically significant difference between the average Canadian traveller and ecotourists as regards many cultural attractions, including museums and art galleries, local festivals and events, and local crafts.

If one is to accept the fact that there currently exists an uneven base with which conceptually and empirically to determine differences in forms or types of nature tourists, it should thus be the task of researchers to focus on the isolation of variables by which to understand better the manifest and latent traits of these groups. Past research (see Bachert 1990) has combined education, experience, and learning into a comprehensive model of wilderness education. In this model learning is a key feature and is a variable which may also provide a basis from which to compare the on-site characteristics of the *nature tourism experience* over other familiar variables such as management, culture, economics and benefits. Other important variables to be considered (besides learning) may include the employment of skills or actions in the pursuit of the primary activity; the degree to which one subscribes to a strong or weak sense of what might be considered as 'biological or preservationist affect' (the emotional tie that one has with plants, animals or nature as a whole); and finally the degree of impact (consumptive to non-consumptive) caused by the type of tourism.

More specifically, learning relates to the need to gain knowledge on-site through interpretation and the information provided by guides and other facilitators. Further research in this area might endeavour to understand the differences or similarities between novelty and curiosity, and learning. In whale-watching, for example, many people just want to see a whale (novelty or curiosity), while conversely others want a more comprehensive learning-based whale-watching experience. In addition, future research may wish to examine the relationship between knowledge and learning in nature tourism. Knowledge can be thought of as information one applies to a situation, whereas learning is something that results from participation. There is little question that nature tourists learn from the experience, but it is important to view learning in terms of the primary motivation of the tourist. Second, skill

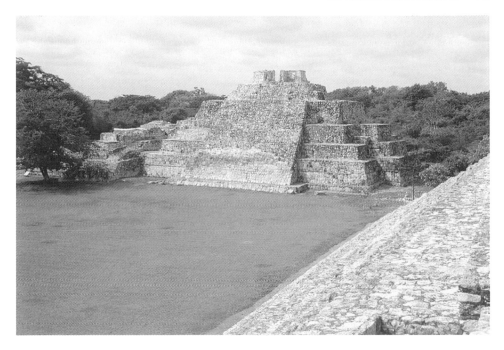

**Plate 2.1** Mayan ruins: major attractions in the peripheral regions of the Yucatán Peninsula, Mexico

**Plate 2.2** To some tourists culture is the primary attraction; to others it is merely a secondary feature of the overall experience

may also be examined because it demonstrates that the application of skill is often employed at the expense of the environment rather than in some other benign fashion. Skill also relates to the previous discussion on primary motivation for participation in the nature tourism activity. There can be a strong case made for the assumption that ecotourism is learning based in its orientation, while other forms of nature tourism (e.g. mountain climbing) may be more skills based in their orientation. Third, biological affect may also provide some inroads in differentiating between nature tourists because it seeks to understand the perceptual place that flora, fauna, geological features, etc. occupy in the mind of the tourist. Consequently, the feelings of such users may dictate the forms of recreation participated in and the effects of such activities on the natural world. This type of thinking relates to the work of Kellert (1985) on the attitudes of people towards animals on the basis of nine variables (negativistic and dominionistic as the most negative attitudes, versus naturalistic and ecologistic as the most positive). Kellert suggests that people and the activities they choose can be correlated with these values towards wildlife. Finally, consumptiveness of the activity (non-consumptive to consumptive) should be considered to illustrate that all forms of outdoor recreation and tourism have some type of impact – however insignificant – on the resource base, with some being less severe than others (consumptive forms of tourism include those that are said to consume resources, such as hunting and fishing).

Despite how the term 'nature tourism' has evolved in the recent past, it is still difficult to equate it with activities that are more consumptive in their orientation (e.g. hunting). The activities that we now group under nature tourism, including adventure tourism, fishing,

## CASE STUDY 2.2

### Nature tourism in the USA

The Recreation Executive report of the US Forest Service (1994) illustrates that nature-based recreation trends as we approach the year 2000 show increases in many non-consumptive outdoor activities. Outdoor photography is up 23 per cent, wildlife watching is up 16 per cent, backpacking is up 34 per cent, and day hiking is up 34 per cent. Also, the National Survey on Recreation and the Environment, a study conducted by the US federal government, concluded that of a number of outdoor recreation activities, birding, hiking, and backpacking experienced the most significant growth in participation between 1982–83 and 1994–95. Birdwatching increased 155 per cent, hiking increased 93 per cent, and backpacking increased 73 per cent. Conversely, hunting and fishing experienced negative growth over this period of time (–12 per cent and –9 per cent, respectively). Further to this, the White House Conference on Tourism in October 1995 indicated the following with respect to the non-consumptive use of birds:

| | |
|---|---|
| Number of participants: | 25 million |
| Retail sales: | $5 billion |
| Wages and salaries: | $4 billion |
| Full and part-time jobs: | 191,000 |
| Tax revenues | |
|    State sales: | $306 million |
|    State income: | $74 million |
|    Federal income: | $516 million |
| Total economic output: | $16 billion |

hunting, whale-watching, and ecotourism might best be labelled 'natural resource-based tourism'. This latter categorisation implies an element of use, which we know corresponds to any form of tourism that occurs outside and which relies specifically on the natural resource base. Furthermore, it is more analogous to the continuum of conservation (saving for use) and preservation (saving from use). Ecotourism, although it entails use, should apply more to the preservation end of this continuum, while hunting and fishing relate more to the aspect of conservation. This sentiment related to the work of Ewert and Shultis (1997), who wrote that 'resource-based' tourism includes a variety of tourism endeavours, including ecotourism, adventure tourism, and indigenous tourism, and their various activities.

## Adventure tourism or ecotourism?

A few hours of mountain climbing turn a villain and a saint into two rather equal creatures. Exhaustion is the shortest way to equality and fraternity – and liberty is added eventually by sleep. – Nietzsche

A close cousin that has developed alongside ecotourism is adventure travel, which in some circles is felt to subsume ecotourism. For example, in suggesting that ecotourism is a branch of adventure tourism, Dyess (1997: 2) admits that he did not have a full appreciation of the heated disagreement that 'exists over the semantics of the two terms, as proponents of ecotourism and adventure travel strive to define and sanctify their own approach to travel'. In a general sense, differentiation between the two forms of travel can simply be based on the type of activity pursued (again, with respect to the primary motivation in participating in the activity). However, in cases where activities are broadly categorised as either ecotourism or adventure tourism, there may be problems, as evident in the following example. Tourism Canada has defined adventure tourism as 'an outdoor leisure activity that takes place in an unusual, exotic, remote or wilderness destination, involves some form of unconventional means of transportation, and tends to be associated with low or high levels of activity' (Canadian Tourism Commission 1995: 5). The following is a list of adventure travel activities developed by the Canadian Tourism Commission (CTC) under this definition: (1) Nature Observation; (2) Wildlife Viewing (e.g. birding, whale-watching); (3) Water Adventure Products (e.g. canoeing, kayaking); (4) Land Adventure Products (e.g. hiking, climbing); (5) Winter Adventure Products (e.g. dog sledding, cross-country skiing); and (6) Air Adventure Products, including hot-air ballooning, hang-gliding, air safaris, bungee jumping, and parachuting. The point to be made from this example is that both ecotourism and adventure tourism products may adhere to the CTC's definition, despite the fact that to many there is a clear distinction between nature observation (ecotourism) and kayaking (adventure tourism). Clearly, further analysis of these activities is required in order to detect similarities and differences. However, while activity is important (an outdoor pursuit) to identifying adventure activities, Priest (1990) writes that there must be an element of uncertainty associated with the event.

The answer to the question of how adventure and non-adventure tourism experiences like ecotourism differ may lie in the realm of social psychology, which examines why participation occurs from a cognitive and behavioural standpoint (instead of from some arbitrary basis established on the setting and other such variables). Both Ewert (1985) and Hall (1992) write that it is risk that plays a primary role in the decision to engage in adventurous activities. In addition, Hall suggests that it is the activity more than the setting that provides the dominant attraction for pursuit of adventure recreation and tourism. Two other factors in analysing one's motivation to engage in risk-related activities include challenge and skill, the fundamental variables behind Csikszentmihalyi's (1990) model of flow (see also Csikszentmihalyi and Csikszentmihalyi 1990). Individuals who are said to have reached a state of flow have matched their personal skills with the challenges of a particular activity. Flow includes a number of main elements as defined by Csikszentmihalyi. These include:

1  *Total immersion into the activity.* This relates to the elimination of distractions that enable the person to lose touch with his or her surroundings.
2  *Enhanced concentration.* A result of the previous factor that allows the participant to forget about the unpleasant tasks that may be associated with the activity.
3  *Actions directed at fulfilling the goal.* The goals and objectives of the event are clearly understood by the participant, who knows how best to approach the situation.
4  *The activity requires skill and challenge.* The relationship between these two variables is important in that if skill far exceeds challenge, boredom will result, whereas if challenge far exceeds skill, anxiety will result.
5  *Flow involves control.* The participant exercises control over his or her movements and the situation, with a degree of anticipation of the events which will unfold.
6  *A sense of transcendentalism.* Here the participant has the experience of transcending his or her physical being, as rooted on the face of the earth, to reach some higher level of understanding or being. A sense of oneness with the surroundings or objects involved in the experience is felt.
7  *The loss of time.* Frequently participants feel as though they have been involved for a short period of time (e.g. one hour), when in fact they have been involved for a long period of time (e.g. four hours).

Csikszentmihalyi (1990) suggests that flow is not necessarily restricted to those who engage in adventurous activities, but also includes those, for example, playing chess or physicians engaged in surgery. According to Hall, then, it may be the desire or the enhanced desire to experience the state of flow that moves individuals to participate in risk or adventure-related activities.

To Quinn (1990), adventure lies deep within the spiritual, emotional, intellectual and objective spheres of humanity, and is the eternal seduction of the hidden (Dufrene 1973: 398). More specifically, Quinn argues that adventure is a desire for a condition that is absent within the individual. Under rather tenuous conditions, the individual must harbour doubt as to the adequacy of his or her ability. The further one attempts to go beyond one's perceived personal talents, the more intense the adventurous situation becomes. It is this element of the unknown that led Einstein to suggest that 'the most beautiful and deepest experience a man can have is the sense of the mysterious' (cited in Mayur 1996: 194). The adventure experience therefore is one that is not discrete but rather one that varies in intensity. The result is that in today's marketplace tourists are able to select from a broad range of hard and soft adventure experiences, offering associated degrees of risk and uncertainty. According to Christiansen (1990), it is the task of the adventure tour operator to provide the client with an adequate perception of risk, while ensuring a high level of safety and security. This is accomplished through an accident-free history, good planning, and the maintenance of the highest level of leadership, skills, and experience. Christiansen provides a good understanding of the difference between soft- and hard-core adventure experiences, as shown in Figure 2.3. The soft adventure activities in the figure are pursued by those interested in a perceived risk and adventure with little actual risk, whereas the hard adventure examples included are known by both the participant and the service provider to have a high level of risk (if, indeed, a service provider is involved in the experience).

In related research the link between motivation and level of experience in mountain climbers was analysed by Ewert (1985), who based his research on the premise that the perception of danger is important in the experience of the risk recreation participant. From

**Figure 2.3** Levels of risk in tour packages

*Source*: Christiansen (1990)

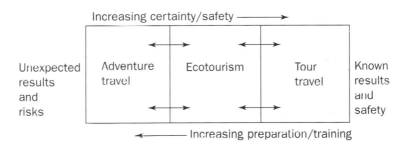

**Figure 2.4** Tourism activity spectrum

*Source*: Fennell and Eagles (1990)

a survey of climbers at Mount Rainier National Park, Washington, he found that inexperienced climbers participated for recognition, escape, and social reasons. Conversely, the experienced climbers were motivated by more intrinsic reasons, including exhilaration, challenge, personal testing, decision-making, and locus of control. The implications of the research are such that over time, climbers who continue with a risk recreation activity such as scuba-diving, mountain climbing and spelunking (cave exploration) may require conditions that are less crowded, more rugged, and less controlled than their novice counterparts.

Fennell and Eagles (1990) focused on the element of risk in examining potential differences between adventure travel, ecotourism, and mass tourism in the framework shown in Figure 2.4. Here it is implied that preparation and training, known/unknown results and risks, and certainty and safety are all variables that may be used to differentiate between these forms of tourism. The figure demonstrates that the three kinds of activity are not mutually exclusive; ecotourism may share some elements of the other two experiences, while still remaining distinct from mass tourism and adventure tourism.

**Figure 2.5** The changing face of ecotourism

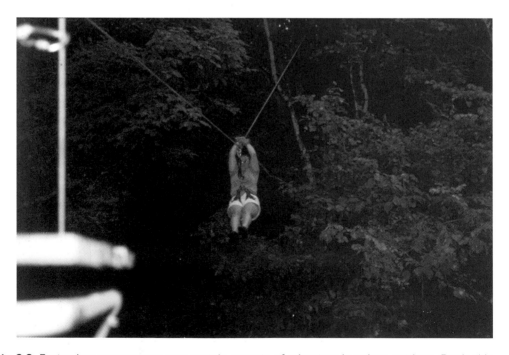

**Plate 2.3** Ecotourism programmes may contain aspects of adventure in unique settings. But is this ecotourism?

Figure 2.5 identifies the evolving relationship between three common AT products, namely ecotourism, adventure tourism and cultural tourism. As suggested earlier, the overlap between these three appears to have become stronger over the past few years, to the point where many studies, usually government or marketing ones (see Canadian Tourism Commission 1995), have considered them as almost completely synonymous (this phenomenon is represented by the acronym ACE in the figure). Depending on the setting and situation, ACE either expands or contracts to represent different concentrations of adventure, culture, and ecotourism, either in name alone or in product content. Those situations that are classed solely as ecotourism avoid the inclusion of conditions that relate to culture tourism or adventure tourism (or both). At the same time, there is the recognition that a continuum exists between hard- and soft-path ecotourism experiences (see Laarman and Durst 1987; Orams 1995). This figure (and the book) attempts to demonstrate that ecotourism should be considered as unique according to its function and role within the tourism marketplace, a stance which is based on the facts that (1) there is little empirical evidence demonstrating homogeneity between adventure tourism, culture tourism, and ecotourism, and (2) there may be an associated dilution factor or effect on ecotourism if these three types of tourism merge into a combined form.

# Tourist typologies

## The general literature

Perhaps as a function of the tremendous demand for different types of travel experiences tourism has blossomed with a rich array of terms describing different types of vacation experiences in different settings, with virtually no research available to describe these various 'types' systematically (Boyd 1991 identifies some 90 types of tourism). Historically, tourism research has tended to concentrate not on tourism types, but rather on tourist types and the various individual traits, characteristics, motivations, needs, and so on of travellers. This has ultimately enabled both researchers and practitioners to better understand tourists on the basis of the types of experiences they seek, as individuals and groups. The following section briefly examines a number of pertinent motivational/behavioural and social/cultural typologies, as a basis for understanding ecotourists.

Christaller quite effectively grasped the notion that, over time, a destination will play host to a rich variety of different types of travellers. He wrote:

> The typical course of development has the following pattern. Painters search out untouched unusual places to paint. Step by step the place develops as a so-called artist colony. Soon a cluster of poets follows, kindred to the painters; then cinema people, gourmets, and the *jeunesse dorée*. The place becomes fashionable and the entrepreneur takes note. The fisherman's cottage, the shelter huts become converted into boarding houses and hotels come on the scene. . . . Only the painters with a commercial inclination who like to do well in business remain; they capitalize on the good name of this former painter's corner and on the gullibility of tourists. More and more townsmen choose this place, now *en vogue* and advertised in the newspapers. . . . At last the tourist agencies come with their package rate travelling parties; now, the indulged public avoids such places. At the same time, in other places the cycle occurs again; more and more places come into fashion, change their type, turn into everybody's tourist haunt.
>
> (Christaller 1963: 103)

While tourist destinations were seen to be transformed under the influence of tourism, so did the type of traveller, either as a result of these changes or as responsible for initiating the changes within the region. Cohen (1972: 172) commented, echoing Christaller, that 'Attractions and facilities which were previously frequented by the local population are gradually abandoned. As Greenwich Village became a tourist attraction, many of the original bohemians moved to the east Village.' Cohen used this as an analogy to demonstrate that travellers were inherently different on the basis of their relationship to both the tourist business establishment and host country. Accordingly, he grouped tourists into four categories, including organised mass tourists, individual mass tourists, explorers, and drifters. This typology reflected a continuum such that the organised mass tourist is seen as the least adventuresome with little motivation to leave the confines of his or her home environmental bubble. The drifter, in contrast, shunned the tourist establishment, searching for the most authentic travel experiences available.

Motivation, or the drive to satisfy inner physiological and psychological needs, has been fundamental to tourism researchers interested in the 'why' of tourist travel. To Pearce (1982), there needs to exist a stronger emphasis in linking roles and motivations with social and environmental preferences of tourists, in order that destinations could be better matched with markets, travel expectations and accommodation. Iso-Ahola (1982) suggested that travel motivation is purely psychological and not sociological in nature. He argued that people travel for basically two reasons: (1) to seek intrinsic rewards (novelty), and (2) to escape their everyday environments (escape). These two motivations may be either of a personal nature (personal troubles or failures) or interpersonal (related to co-workers, family, or neighbours). The amalgam of these elements is a four-cell matrix, where a single tourist could theoretically go through one or all aspects during the course of one trip.

MacCannell (1989) also considered the fundamental differences between traveller types in examining the social structure of tourist space based on 'front' and 'back' regions. Front regions are those readily experienced by tourists and places where hosts and guests regularly interact. Conversely, back regions are the preserve of the host and are essentially non-tourism oriented in their function. Tourists in search of authenticity penetrate back regions in the hope of acquiring real day-to-day mannerisms of residents. To what degree tourists are willing or able to penetrate back regions – in identified access zones of tourism areas – may be important to their achieving their overall purpose. Although these 'places' are actual locations in the conventional economic/geographical context, in the mind of the tourist or tourist group they hold special value in defining the travel experience. In general, these back regions are sociological.

## Profiling the ecotourist

Some of the earliest studies on ecotourism attempted to classify ecotourists on the basis of setting, experience and group dynamics. For example, Kusler (1991) typified ecotourists as belonging to three main groups, including:

1   *Do-it-yourself ecotourists.* Despite their relative anonymity, this group comprise the largest percentage of all ecotourists. These individuals stay in a variety of different types of accommodations, and have the mobility to visit any number of settings. Their experience, therefore, is marked by a high degree of flexibility.
2   *Ecotourists on tours.* This group expects a high degree of organisation within their tour, and travel to exotic destinations (e.g. Antarctica).
3   *School groups or scientific groups.* This group often become involved in scientific research of an organisation or individual, often stay in the same region for extensive periods of time, and are willing to endure harsher site conditions than other ecotourists.

Conversely, Lindberg (1991: 3) emphasises the importance of dedication and time as a function of defining different types of ecotourists, including what tourists wish to experience from ecotourism, where they wish to travel, and how they wish to travel. Lindberg identified four basic types:

1   *Hard-core nature tourists*: scientific researchers or members of tours specifically designed for education, removal of litter, or similar purposes;
2   *Dedicated nature tourists*: people who take trips specifically to see protected areas and who want to understand local natural and cultural history;
3   *Mainstream nature tourists*: people who visit the Amazon, the Rwandan gorilla park, or other destinations primarily to take an unusual trip; and
4   *Casual nature tourists*: people who experience nature incidentally as part of a broader trip.

Consistent with the increase in ecotourism opportunities, and the associated typologies of tourists travelling on such trips, has been an increase in the volume of literature available on the ecotourist, with much of it focusing on presenting a profile of this tourist type. Initial studies decidedly pointed to the fact that the ecotourist was predominantly male, well educated, wealthy, and long-staying. For example, Wilson (1987: 21) reported the following in researching 62 tourists visiting Ecuador:

The male/female ratio was 52% to 48%, and the mean average age was 42. . . . Twenty-seven percent earned a family income between US $30,000 to $60,000, before taxes annually. Approximately one-quarter earned more than $90,000 per year. About 30% had bachelors degrees, and a little over 10% had doctoral degrees.

Both Fennell and Smale (1992) and Reingold (1993) report similar results in their work on Canadian ecotourists. On average, the Canadian ecotourists in the Fennell and Smale study were 54 years of age, with the majority in the 60–69 age cohort. The sample was predominantly male (55 per cent), earned on average about CDN$60,000, with almost one-third and two-thirds combined having undergraduate and graduate degrees, respectively. According to these authors, this education is well above the national average of about 19 per cent and 4 per cent of Canadians having a bachelor's degree and graduate training, respectively. In Reingold's work, 24 per cent of Canadian ecotourists were 55 to 64 years of age, 36 per cent had annual incomes over $70,000, 65 per cent had a university degree; however, 64 per cent of the respondents of this study were female. This latter statistic goes against much of the previous research suggesting that males are slightly more representative of this group. The general outdoor recreation literature supports the fact that most participants – two-thirds or even three-quarters – are of the male gender (Hendee *et al.* 1990).

The studies involving ecotourists have been found to mirror related research on birdwatchers and wildlife tourists. For example, Applegate and Clark (1987) report that more men than women birdwatch and that they have strikingly high levels of affluence and education with more than 50 per cent of the respondents of their study completing four years of college. The findings of Kellert (1985) indicate that committed birders were 73 per cent male with an average age of 42. Committed birders were also far better educated (nearly two-thirds had college and/or graduate school education) and had higher incomes than respondents who did not birdwatch. Polar-bear-viewing tourists to the Churchill region of Manitoba, Canada, were found to be middle-aged to older aged, well educated and financially successful (see MacKay and McIlraith 1997). In a more recent study of this region, Lemelin *et al.* (2002) found that these tourists numbered slightly more as males than females, 44.4 per cent were aged 45–64, 40 per cent earned in excess of

$100,000 as a household income, and 88 per cent had a post secondary degree or diploma. As more studies of this nature surface in the future it will be interesting to see how the profile of the ecotourist changes (or stays the same) as a function of many variables, including the types of experiences and products offered to ecotourists, and the maturity of the industry.

Beyond an analysis of socio-demographics, Fennell and Smale (1992) directly compared the average Canadian tourist against a sample of ecotourists (based on the work of Fennell 1990). These researchers used the results of a 1983 Canadian study (Tourism Attitude and Motivation Study (CTAMS)) that analysed the general attitudes and benefits sought by a sample of 11,501 Canadian tourists. The questions used in the CTAMS study were duplicated and applied to a sample of ecotourists who had recently returned from a Costa Rican ecotourism trip between 1988 and 1989. Of 98 surveys mailed to the ecotourists (tourists were contacted by obtaining mailing lists of Canadian ecotour operators offering programmes in Costa Rica), 77 were returned and usable. The results of this study are found in Table 2.2.

In general, the findings illustrate that the benefits sought by ecotourists may be found in new, active, and adventuresome activities and involvements, while the Canadian population benefits were more strongly related to sedentary activity and family-related endeavours. Ecotourists were found to pursue attractions related to the outdoors (wilderness areas, parks and protected areas, and rural areas), while attractions related to cities and resorts were more important to the average Canadian traveller. The implications of the research are such that they empirically demonstrate that there is indeed a difference between ecotourists and general travellers in terms of their trip-related needs and focus. Kretchman and Eagles (1990) and Williacy and Eagles (1990) followed the initial survey design of Fennell (1990) in comparing ecotourists in a variety of different settings. These studies were combined by Eagles (1992) in a comprehensive overview of the travel motivations of Canadian ecotourists. His results generally substantiate the results of Fennell and Smale (1992), above, in suggesting that ecotourists are fundamentally different, in their travel motivations, from the average traveller.

As suggested earlier, an effective way in which to differentiate between types of tourists visiting particular regions is through an examination of various personal characteristics such as primary motives, benefits sought, and so on. Turnbull (1981), however, provides an interesting argument against such an approach. He suggests that although on the surface nature tourists travelling to Africa state that it is primarily the animals that bring them to the game parks of East Africa, on the basis of his observations he feels that the reasons for visiting Africa run much deeper. As an anthropologist, Turnbull believes that perhaps latently it is also the 'Africa' and the 'Kilimanjaro' that bring such tourists to this setting. In essence, he believes that people want to tap into their distant past when humans had a much stronger relationship with animals, seeing them as something more than just prey to be hunted, and it is to experience something of this relationship that is really desired by present-day tourists. He states that the tourists

> expect an indivisible, natural whole made up of both human and animal components. But unless they are unusually lucky this is not what tourists find on the organised safari. All that is observed is man-made: game parks devoid of the herders and hunters who used to live there as an indispensable part of the ecosystem.
>
> (Turnbull 1981: 34)

This theory gives us another means by which to further examine the intergroup differences that exist between various forms of tourism (i.e. ecotourists and non-ecotourists), but also the intragroup differences (i.e. soft-path ecotourists and hard-path ecotourists) as well. We

Table 2.2 *Relative importance of selected attractions and benefits to Canadian travellers and ecotourists[a]*

| Variable | General population | | Canadian ecotourists | |
|---|---|---|---|---|
| | Mean[b] | s.d. | Mean | s.d. |
| *Important benefits to ecotourists[c]* | | | | |
| Experiencing new and different lifestyles | 2.67 | 1.02 | 1.95 | 0.78 |
| Trying new foods | 2.85 | 0.99 | 2.43 | 0.82 |
| Being physically active | 2.36 | 1.03 | 1.88 | 0.92 |
| Visiting historical places | 2.71 | 1.04 | 2.35 | 0.74 |
| Seeing as much as possible | 2.01 | 0.99 | 1.77 | 0.81 |
| Being daring and adventuresome | 2.91 | 0.98 | 2.62 | 1.01 |
| Meeting people with similar interests | 2.08 | 0.94 | 1.91 | 0.71 |
| | | | | |
| *Important benefits to population[c]* | | | | |
| Watching sports | 3.16 | 0.96 | 3.84 | 0.37 |
| Visiting friends and relatives | 1.83 | 1.03 | 3.25 | 0.82 |
| Doing nothing at all | 2.86 | 1.00 | 3.72 | 0.60 |
| Being together as a family | 1.66 | 0.98 | 2.92 | 1.04 |
| Reliving past good times | 2.46 | 1.06 | 3.36 | 0.79 |
| Visiting places my family came from | 2.70 | 1.16 | 3.45 | 0.84 |
| Feeling at home away from home | 1.81 | 0.91 | 2.64 | 0.93 |
| Having fun and being entertained | 1.95 | 0.88 | 2.72 | 0.94 |
| | | | | |
| *Important attractions to ecotourists[c]* | | | | |
| Wilderness areas | 2.34 | 1.09 | 1.06 | 0.37 |
| National parks and reserves | 2.21 | 1.01 | 1.14 | 0.35 |
| Rural areas | 2.34 | 0.94 | 1.49 | 0.60 |
| Mountains | 2.34 | 1.07 | 1.50 | 0.66 |
| Lakes and streams | 2.05 | 0.99 | 1.57 | 0.59 |
| Historic sites and parks | 2.37 | 1.01 | 2.05 | 0.81 |
| Cultural activities | 2.66 | 0.97 | 2.32 | 0.87 |
| Oceanside | 2.15 | 1.07 | 1.97 | 0.78 |
| | | | | |
| *Important attractions to population[c]* | | | | |
| Indoor sports | 2.98 | 0.94 | 3.85 | 0.35 |
| Amusement and theme parks | 2.74 | 0.99 | 3.80 | 0.54 |
| Nightlife and entertainment | 2.72 | 1.04 | 3.70 | 0.56 |
| Gambling | 3.61 | 0.69 | 3.96 | 0.34 |
| Shopping | 2.45 | 1.01 | 3.14 | 0.74 |
| Resort areas | 2.56 | 1.01 | 3.26 | 0.82 |
| Big cities | 2.94 | 0.93 | 3.39 | 0.69 |
| Beaches for swimming/sunning | 2.34 | 1.10 | 2.79 | 0.96 |
| Predictable weather | 2.11 | 0.97 | 2.40 | 0.81 |
| Live theatre and musicals | 2.99 | 1.01 | 3.21 | 0.91 |

Source: Fennell and Smale (1992)
Notes:
[a] Differences between general population and ecotourists are statistically significant at 0.05 level
[b] Scale: 1 = 'very important'; 4 = 'not at all important'
[c] Attractions/benefits are in descending rank order based on magnitude of difference between groups

cannot blindly accept the notion that ecotourists are one homogeneous group; instead, as outlined above, they may be differentiated on the basis of many variables.

## Conclusion

Despite the volume of literature that has emerged on the topic of ecotourism, it is clearly at an early stage of development, as is our understanding of how ecotourism differs from other types of tourism. More research is needed to help formulate definitions acceptable to those working in the field, to help overcome this critical absence of focus. The literature points to the fact that ecotourism is one aspect of nature-oriented tourism, which includes many other types of tourism and outdoor recreation, both consumptive and non-consumptive. Ecotourism is said to relate to adventure travel only inasmuch as it shares similar environmental settings. Differentiation between these two forms of tourism must be done on the basis of social psychology. Finally, more studies are needed to further our understanding of who the ecotourist is, especially in relation to other types of tourists.

## Summary questions

1 Many have thought that ecotourism is a form of tourism which came about in the early 1980s. Why is this a contentious issue?
2 Linking with Chapter 1, how are sustainable tourism and ecotourism related? How are they different?
3 Why do you think that it has been so difficult to come up with one widely accepted definition of ecotourism?
4 What are the advantages and disadvantages of ACE tourism in providing the means by which to differentiate between cultural tourism, adventure tourism and ecotourism?
5 What are some of the characteristics which have been used to differentiate ecotourists from other types of tourists?

# 3 Natural resources, conservation and protected areas

> We abuse land because we regard it as a commodity belonging to us. When we see land as a community to which we belong, we may begin to use it with love and respect. – Aldo Leopold

The above quotation aptly describes the dilemma that has existed for some time in humanity's struggle to balance natural resource utilisation with preservation. Globally, we have been slow to recognise how our economic development activities and patterns of consumption have come to modify the resource base, which is a legacy of domination that is firmly entrenched within Western society. However, although the implications of our socio-economic actions are debated, we are decidedly ambivalent on how best to temper development-oriented activities. In this chapter, a general discussion of resources, conservationism, and parks is used to describe the polarisation of thought on the value and role of resources within society. The purpose is to provide a backdrop from which to venture off into a more specific discussion of ecotourism as it relates to economics, parks, and so on. An attempt is made to highlight some of the main considerations of the development of a conservation philosophy in light of the pressures that humans have placed, and will continue to place, on the natural world. Parks and protected areas are discussed as settings that have an important role in balancing ecological integrity and touristic demand.

## The exploitation of the natural world

### Natural resources

People use resources to accomplish ends in a variety of work and leisure settings. In the workplace, individuals make use of other people (human resources) as well as facilities and equipment (physical resources), and in some cases the natural environment (natural resources), to accomplish their various tasks. During travel the same holds true, but the motivational focus alters to incorporate the use of such resources, independently or in combination, to satisfy a range of personal desires.

Zimmerman (1951: 15) noted that resources 'are not, they become; they are not static but expand and contract in response to human actions'. That is, elements of the resource base such as trees, water, rocks, etc. do not become resources until they are capable of satisfying human needs. Zimmerman referred to elements of the environment that do not have human utility as 'neutral stuff'. The further humanity delves into the realm of the natural world, the more this neutral stuff becomes transformed into resources. Culture also has an effect on who uses such resources, and to what degree. For example, oil becomes a resource if people endeavour to develop the knowledge and ability to extract it, and combine it with the technology required to build the implements we use for work, leisure and survival (cars, furnaces, and so on). Resources, therefore, are dynamic both in space and time and are very much related to the perception of their worth to a particular person or society, as suggested by Mitchell (1989: 2): 'Natural resources are defined by human perceptions and attitudes, wants, technological skills, legal, financial and institutional arrangements, as well as by political systems. . . . Resources, to use Zimmerman's words, are subjective, relative and functional.'

The pursuit of touristic needs occurs along a broad physical site development continuum, from those settings that have been substantially modified by humans to pristine environments with very little human intervention. Chubb and Chubb (1981) suggest that the dividing line between what is developed and what is undeveloped is contingent upon one's perception of the meaning of the word 'developed' in relation to the tourism setting. Developed resources include highways, facilities, sewerage, buildings, and so on, that facilitate the use of a given area. Conversely, undeveloped resources may be found both in urban and wilderness environments but the degree to which they are recognised as such is individual dependent and perhaps situation dependent. These authors outline seven different types of undeveloped resources as they apply to outdoor recreation (and hence tourism). Although independent, each of the following must be envisioned as integrated, as it is the combination of elements of the environment that structure our tourism experiences.

- *Geographic location*. The characteristics of space that determine the conditions, in association with other variables, for participation (e.g. skiing).
- *Climate and weather*. Determined by latitude and elevation relative to large landforms, mountains, ocean currents, and high-altitude air currents. Along with geology, climate is the prime controller of the physical environment, affecting soils, vegetation, animals and the operation of geomorphological processes such as ice and wind.
- *Topography and landforms*. The general shape of the surface of the earth (topography) and the surface structures that make some geographical areas unique (landforms). A landform region is a section of the earth's surface characterised by a great deal of homogeneity among types of landforms.
- *Surface materials*. The nature of the materials making up the earth's surface, including rocks, sand, fossils, minerals, soil, sand, etc.
- *Water*. This substance plays a critical role in determining the type and level of outdoor recreational participation in ocean and sea environments as well as freshwater settings (lakes, rivers, and wetlands).
- *Vegetation*. Vegetation refers to the total plant life or cover in an area. Recreation quite often is dependent on plant life directly (tourists taking pictures of unique plant species), or indirectly (trees acting as a wind barrier for skiers).
- *Fauna*. Animals can play a significant role for recreational activities that are both consumptive and non-consumptive in nature. Forms of consumptive recreation view wildlife from the notion that animals have a utilitarian or dominionistic function (e.g. fishing, hunting, etc.). Non-consumptive recreation on the other hand has a softer impact on the resource base (e.g. birdwatching).

It should be noted that these resources may act either as catalysts in facilitating and drawing people to a tourist region or as constraints to visitation. A recent case in point is Montserrat. This small island state in the Caribbean is blessed with an abundance of natural features in its 102 km², and these usually act as a catalyst to draw people to this destination. Given its relative abundance of natural and climatic features, Montserrat has been referred to as an excellent example of a region where ecotourism may prosper (Reynolds 1992; Weaver 1995). However, in 1997 the tourism industry took a turn for the worse when the island's dormant volcano erupted, leaving the island, and the island's economy, in a critical state. Natural disasters such as this have not only an immediate effect but also a long-term effect, as Weaver (1995: 601) writes in anticipating the fate of Montserrat's tourism industry:

> The inevitable hurricanes and earthquakes of the future will periodically curtail ecotourism by damaging not only the physical environment, but the roads, hiking trails, and viewing platforms upon which the sector depends. Furthermore, a prolonged moratorium on tourist activity may be necessary in some areas as the environment recovers from a natural disaster.

## The earth's bounty

In an influential treatise on the historical roots of our present-day ecological dilemma, White (1971) argues strongly for the notion that, where formerly humankind was part of nature, over the past several hundred years our species has become the exploiter of nature and natural resources. The strength behind this conviction lies in how Western Christianity and capitalism have polarised humankind and nature by insisting that 'it is God's will that man exploit nature for his proper ends' (White 1971: 12). Historical accounts charge that many scientists of the Middle Ages and Renaissance periods were tempered by this Christian philosophy. During the great scientific renewals of the 1500s and 1600s, a mechanistic and static perception of nature developed, replacing the passive mystic and organic views of earlier times. The push was towards a type of science that had designs on enlarging the bounds of the human empire in the endless pursuit of knowledge and control in order to 'regain the level of understanding and power once enjoyed in the Garden of Eden' (Bowler 1993: 85).

Another significant element that contributed to humankind's dominance over nature was fear of the unknown. Although scientists in medieval times had begun to unravel some of nature's mysteries, their lack of understanding of the intricacies and interconnectedness of the environment contributed to the feeling that marginal places – such as the wilderness – were areas that had to be subdued (Short 1991). Land and animals were dichotomised as either settled or savage, cultivated or uncultivated, domesticated or wild. Through Christian eyes, wilderness areas (forests, bogs, and the like) were settings that contained pagans exercising pagan rites. In order to bring about religious order, such lands had to be cut down and cleansed. The abundant body of literature that exists on human–wilderness relations illustrates that there is no specific material object or space that one can identify as being 'wilderness' (Nash 1982). Wilderness is a concept that produces specific feelings or moods within people, and occurs within one's mind as a perceived place. Such a place may be wilderness to one, but not to another. This attitude prevailed in the Puritan mind as the first settlers set foot on North American soil. The new settlers were religiously and morally obligated to spend the bulk of their time in work-related activities. Leisure was severely curtailed and strictly regulated in accordance with religious protocol. The implications for nature were such that the wilderness represented both a challenge and an obstacle to a sustained European colony within the New World.

Socio-economically, significant changes were occurring in Britain during the period 1750 to 1850. It was a time of massive transformation and development of British industry by the use of new technologies and the circuit of industrial capital, which inevitably contributed to drastic changes in the leisure patterns of British citizens (a rural-to-urban population shift, longer working hours, and the exploitation of workers). Emerging within Britain and later in North America during the early 1800s was a progressive human attitude geared towards development that emphasised the following (Short 1991: 1): the creation of liveable places and usable spaces (houses, industry); (2) the regarding of wilderness areas as waste and desolation; (3) the idea that civilised human society gave significance to the world; and (4) the conquest of wilderness as a signal of human achievement.

---

## CASE STUDY 3.1

### *Natural history travel in Shetland, Scotland*

Butler (1985) writes that up to 1750 the Highlands and Islands of Scotland were virtually *terra incognita* to the people of the rest of Britain. The Shetland Islands were no exception. However, as a consequence of its unique location, Shetland provided both a geographical and a cultural link between Scandinavia to the east and Great Britain to the south. Flinn (1989) suggests that before 1850 the people who travelled to Shetland were artists, geologists, naturalists, physicists, and surveyors. In 1814 Sir Walter Scott visited Shetland and used the region as the setting for his novel *The Pirate*. As Simpson (1983) illustrates, Scott's novels suffused a romance and drama of the book's region in the minds of the readers, compelling them to visit such places. In 1832 an ironmaster and naturalist, G.C. Atkinson, travelled to Shetland, perhaps as a result of Scott's influence. He wrote, 'I have long felt the greatest interest in descriptions of novel and extraordinary scenery and of the inhabitants and natural productions of regions that have been little known, either from their difficulty in attainment . . . or from their being so near.' By 1850, Flinn contends, the nature of the Shetland visitor was somewhat different. He wrote, 'by 1859 Shetland had a frequent steamer service from the south during the summer months, and a trunk road system which made the islands more accessible to tourists of a less hardy and enterprising breed' (ibid.: 235). Tourism, it appears, had displaced the traveller bent on exploration and challenge with a breed less willing to endure the hardships of travel.

---

## The roots of conservation

As the modification of the natural world began to intensify, it was beginning to be understood in France and Britain that deforestation was a predominant cause of soil erosion and poor productivity. These discoveries prompted early attempts at conservation in Britain, and in British India, where forested lands were set aside for the purpose of shipbuilding. Such areas were referred to as *conservancies*, and the foresters in charge of these areas as *conservators* (Pinchot 1947). There is evidence, however, to suggest that reforestation policies were being implemented in Britain in the seventeenth century (Bowler 1993).

In North America, conservationism evolved on three fronts (Ortolano 1984). The first involved the view that conservation should entail the maintenance of *harmony* between humankind and nature, the second that conservation related to the *efficient* use of resources. The final perception was that conservation – preservation – could ideally be attained from the standpoint of religion and *spirituality*. This latter view of conservation engendered a philosophy with the aim of saving resources *from* use rather than saving them *for* use (Passmore 1974). Each of these perspectives is discussed in further detail below.

## Harmony

In the United States, a former Minister to Turkey and key founder of the Smithsonian Institution, George Perkins Marsh, became an instrumental figure in illustrating to Americans that their actions (commerce and lifestyles) were uniquely potent. Marsh wrote that:

> The earth is fast becoming an unfit home for its noblest inhabitant, and another era of equal human crime and human improvidence . . . would reduce it to such a condition of impoverished productiveness, of shattered surface, of climatic excess, as to threaten the depravation, barbarism, and perhaps even extinction of the species.
>
> (Bowler 1993: 319)

His *Man and Nature*, originally published in 1864, recognised that in the changing conditions of nineteenth century America, harmony between human influences (modifications) and the natural world could be achieved only through society's commitment to a moral and social responsibility to future generations. (This view was rekindled over eighty years later by the land ethic philosophies of Aldo Leopold.) Marsh identified the collective power of large corporations as the factor most responsible for the increasing negative impacts to the natural world. Politically, the accountability of the large firms did not become an issue until the turn of the century. The Roosevelt administration recognised that the federal government had been too generous in granting favours (leases, land rights, etc.) to such corporations (Hays 1959), and as a result attempted (1) to find an adequate means of controlling and regulating corporate activities; and (2) to resist the efforts of corporations to exploit the natural resources of the nation on their own behalf.

At the time, changing technology, industrialisation, urbanisation, population growth, and transportation all contributed to the feeling that the American frontier was diminishing. This fact was most emphatically illustrated by the direct relationship between the advancement of the transcontinental railroad and the decline of American buffalo. The combination of a loss of habitat and the shooting of buffalo for sport (from the trains themselves) contributed significantly to the downfall of this species. Space was also a critical factor in the frontier mentality. Americans had reached the supposed limits of their 'manifest destiny' by encountering the Pacific to the west, Canada to the north, and Mexico to the south. In Canada, the frontier attitude lingered on because of the challenges encountered in settling and harnessing the conditions of the north.

## Efficient use

By the beginning of the twentieth century, Americans were ripe to exercise concern over the fate of their resource base (Nash 1982). The concept of conservation became a vehicle to represent the new frontier; the means by which American society could maintain vitality and prosperity. However, despite conservationism's apparent simplicity – the wise use of natural resources – fierce debate surfaced over how such resources ought to be utilised, if at all.

The efficient use of resources perspective represented the opposite end of the conservation spectrum from the perspective of preservationism, and was championed by an American by the name of Gifford Pinchot. Pinchot developed a concept of natural resources for the greatest good of the greatest number for the longest time (Herfindahl 1961). More directly, conservation was to engender direct control over natural resources on the basis of three principles (Pinchot 1910): (1) to develop the continent's existing natural resources for the benefit of the people who live there now; (2) to prevent the waste of natural resources; and

(3) to develop and preserve resources for the benefit of the many, and not merely for the profit of a few.

This philosophy of conservation, at the very least, recognised the need for people to acknowledge that natural resources were finite. Industry and the individual both had to be more accountable for their actions under the rationale that what was utilised today might have repercussions for those in the future. From this standpoint, conservationism was indeed progressive, as suggested by Hays (1959: 264):

> The broader significance of the conservation movement stemmed from the role it played in a transformation of a decentralised, non-technical, loosely organised society, where waste and inefficiency ran rampant, into a highly organised, technical, and centrally planned and directed social organisation which could meet a complex world with efficiency and purpose.

## Spirituality

A more romantic view of wilderness was developing in response to the technological and industrialised transformation of Britain and Europe. Based on the works of Erasmus Darwin, Wordsworth, Coleridge, and Carlyle, Romanticism developed from the more regressive view that society had declined from past, more harmonious times. Romanticism embodied a deeper spirituality and awareness that a simpler life was attainable without the complications of a society blemished by materialism, and could be accomplished under the following conditions (Short 1991): (1) untouched spaces had the greatest significance; (2) these spaces had a purity which human contact degrades; (3) wilderness was a place of deep spiritual significance; and (4) the conquest of nature was a fall from grace.

The first proponent of the Romantic philosophy within North American society was Ralph Waldo Emerson, who had met and been inspired by the romantic poets of Britain in the early 1830s. The main doctrine of Emerson's interpretation of Romanticism surrounded the belief that although mankind was firmly rooted to the physical world, people had the ability to 'transcend' this condition (spiritually) in searching for and achieving deeper philosophical truths. Emerson's transcendentalism was a spiritual doctrine of humankind and nature. To Emerson, humankind was divided into materialists and idealists, with the first class founded upon experience and the second on consciousness. The materialists were to insist on facts, history, circumstances, and animal instincts; the idealists on the power of thought, will, and miracle. His major work, *Nature* (1835), had a significant impact on many other writers to follow, including Henry David Thoreau, Herman Melville, John Burroughs, Walt Whitman, and John Muir.

Nature in North American society thus began to hold and inspire a small but articulate core of advocates paving a clear path out of the downward-spiralling pattern of a materialistic, consumptive society. Although the works of Emerson were largely idealistic, the transcendentalist movement also provided the rationale for practical change within American society. Thoreau, for example, campaigned to have the US government establish national preserves for the purpose of ensuring the future well-being of animals (Finch and Elder 1990). Such a call was well ahead of its time, as the United States did not endeavour to establish protected areas until a number of years later.

The emergence of a second tier of conservation (environmentalism or the green movement) occurred in the 1960s in response to the rapid and overwhelming increase in the impact of technology in society. Publications such as Rachel Carson's *Silent Spring* were effective in calling attention to the insidious effects of chemical use within society. In addition, ecology and the growing importance of science had begun to become institutionalised and gain acceptance as a mechanism to evaluate many social and ecological ills of the time. Bowler (1993) writes that the most militant supporters of the environmental movement opposed the entire economic structure of society in favour of a reversal to a simpler, more natural state. Examples of organisations/movements advocating a more radical approach to environmentalism include Greenpeace, Deep Ecology (founded by Arne Naess), and the Eco-Feminist Connection, which is believed to go deeper than Deep Ecology in addressing the man-centred approach to development within society and the fact that human females best represent the close tie with Mother Nature. Those environmentalists considered less militant called for the protection of selected parcels of land that held natural and cultural significance, with the request that industrialisation and development occur outside the realm of such areas. According to Bowler, a fundamental difference existed between the two camps. Those less enthusiastic (i.e. those in favour of the development of nature reserves) recognised that minor changes could occur within the current system, while the most enthusiastic environmentalists wanted to destroy the existing social order. It is the former group that, in a practical sense, has made the most significant gains in the context of the global arena. Ecotourism, to some, is merely an extension of this philosophy of 'working within the system' and one that, at least conceptually, attempts to knit the elements of economy and ecology together (via parks) through the tenets of environmentalism and sustainable development (see Chapter 1).

## Parks and protected areas

The concept of 'park' is one that is firmly established within civilisation (Smith 1990a; Wright 1983). The Greeks and Romans met at designated open spaces (*agorae*), while in medieval times the European nobility used their private lands as hunting reserves. With very few unmodified open spaces left in Britain by the nineteenth century, the concept of 'park' began to be recognised as a means by which to secure outdoor recreation opportunities in the countryside. To accommodate demand, the nobility began to lease open space for summer and winter recreation of all classes within British society. The importance of parks was more formally established by the Municipal Corporations Act in 1835, allowing for the creation of municipal parks and for the secure right of public recreation.

In Britain, the parks movement stemmed largely from the response of the British public to the effects of urbanisation, pollution, and loss of leisure equated with the Industrial Revolution. The same could not be said for the evolution of parks in Canadian society. Wright (1983: 45) illustrates that urban parks in Canada were created to satisfy a concept rather than a reality:

> the establishment of local parks in the 1850–1880 period in Ontario was merely the continuation in Canada of a program and philosophy adopted in Britain and maintained by the elite in their own environment, not because the conditions in Canada demanded its implementation but because the colonial settlers wished to preserve the values and beliefs inherited from their ancestral homes.

Ontario was the first of the provinces in Canada to enact legislation governing municipal park development in 1883 (Eagles 1993). This occurred at a time when there was a heightened awareness in North America of the need for parkland. The world's first national park had been created in Yellowstone in 1872, and Canada's first national park was to be created two years after Ontario's municipal legislation in 1885.

Parks and protected areas (the terminology used in this section to refer to public lands held in trust with both a recreation/tourism and conservation/preservation mandate, and owned and usually operated by a public agency) have a certain mystique to travellers interested in some of the best representative natural regions or countries. In fact, it was Johst (1982) who suggested that visitation to parks may increase by virtue of their designation as such. Simply stated: parks and protected areas often times generate more recreational use simply because they are recognised as parks (although McCool (1985) contests this point).

Harroy (1974) has shown that Yellowstone was created to satisfy a broad mandate of concerns that had emerged from the United States' frontier mentality. Foremost, the park was set up to prevent the exploitation of wildlife and the environment, for the purpose of recreation, and finally as a means of scientific study. Canada's first national park, Banff, on the other hand, was established for political and economic reasons, including the generation of tourism dollars (largely from the therapeutic and recreational benefits of the hot springs) for the purpose of offsetting the cost of building the transcontinental railway (Lothian 1987). Banff's popularity quickly escalated to the point where the park was absorbing a wide variety of recreational demands including fishing, horseback riding, hunting, mountain climbing, and so on, but also other resource demands such as mining, logging, grazing, and a town site. Rollins (1993) suggests that as a case study, Banff represented the types of problems that Parks Canada has had to address over the past 110 years regarding the management of natural features, with many conflicting stakeholders harbouring consumptive and non-consumptive designs on the use of park resources.

In Canada and many countries around the world, national parks are broadly mandated with the dual purposes of protecting representative natural areas of significance, and encouraging public understanding, appreciation, and enjoyment. Their main focus, therefore, is to balance recreational use with the protection of unique physiographic land and water regions. Historically, the preservation ideal within parks was not fully developed or emphasised. However, as the system of protected areas continues to grow (as illustrated by the increasing circles over time in Figure 3.1), park management philosophies have become better integrated, recognising that parks do not exist as ecological islands, but must be managed according to environmental conditions both inside and outside their boundaries (Dearden 1991).

The debate surrounding the threats to parks and park management has prompted some researchers to acknowledge a situation of crisis in the national parks in the United States due to both internal and external factors (Chase 1987), and even that national parks in their current form are not sufficient to remain effective in the future (Janzen cited in Chase 1989). Janzen illustrates, using the African game parks as an example, that the establishment of wilderness lands has been self-defeating in that they slowly become consumed by poaching and adjacent farmers. To Janzen, a preferable approach would be to view the national park as an agent of social change, where local people would be stakeholders within the park and their social and economic activities (e.g. tourism and farming) would be ecologically sustainable.

Lovejoy (1992) writes that today, parks serve a variety of purposes but also face a number of pressures. He cites fertiliser use infiltrating the park environment from adjacent lands, the temptation of local people to use park resources, self-serving political interests that influence park management, overpopulation, and habitat fragmentation as frequent dilemmas challenging park managers. Worldwide, though, Lovejoy suggests that increased tourism visitation stands as one of the most persistent problems facing parks and protected areas. He identified Yosemite and Yellowstone National Parks in the United States as examples particularly representative of this problem. Yosemite is overused during the summer, and the Yellowstone authorities, despite the parks integration with surrounding communities, were criticised – by those with business enterprises in or around the park – for allowing the extensive fires of 1989 to burn, which in turn affected the tourism industry. Parks, therefore, have come to rely on tourism as a means by which to generate income from a growing world

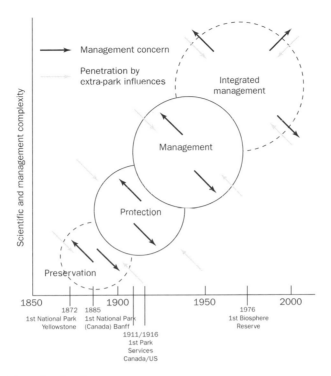

**Figure 3.1** The evolving role of parks

*Source*: Dearden and Rollins (1993)

population with increasing disposable time, financial well-being, and personal mobility (see also Chapter 6).

The problems between parks and local populations outlined above are mirrored by Hough (1988), who feels that restrictions of access to traditionally used resources, and the disruption of local cultures and economies by tourists, etc., have led to hostility, resentment, and damage to park property. He identifies eight key obstacles to effective management of park–people relationships, namely the institutional structure of national parks (i.e. the concept of parks and their policy structures); the lack of trust between local people and park authorities; the lack of communication between parks and local people; the large number of different stakeholders in the park; the polarisation of power between government and local populations; the risk and uncertainty in entering into discussions aimed at reducing conflict; the enforcement of agreements between the park and local population; and the lack of opportunity for all to participate in the decision-making process.

In general, the threats to parks and parks management have, according to Dearden and Rollins (1993), evolved over time, having been primarily internal but now more external in their orientation. This evolution has coincided with the belief that the roles of parks have changed significantly from when they were developed initially (at least in Canada), at which time the primary role was to cater to recreationists. This is no longer the case, at least in theory, however, with the recent changes to the Canadian Parks Act in 1988 to underscore the importance of ecological functioning of parks above the recreational component.

Section 5(1.2) of the 1988 amendments suggests that maintenance of ecological integrity through the protection of natural resources is to be the first priority when considering park zoning and visitor use in the management plan (Canada, Parliament 1993). Park zoning is one of the key planning and management tools within parks. Zones are established on the basis of natural resources and their need for protection, and capacity to absorb recreational involvement. An overview of the theoretical basis for zoning in national parks in Canada (Environment Canada 1990) is illustrated as follows (for more of a history on the development of outdoor recreation land classification, the reader should consult the US Outdoor Recreation Resource Review Commission reports of the early 1960s):

- *Zone 1: Special Preservation.* Specifies areas or features which deserve special preservation because they contain or support unique, rare, or endangered features or the best examples of natural features. Access and use will be strictly controlled or may be prohibited altogether. No motorised access or human-made facilities will be permitted.
- *Zone 2: Wilderness.* Extensive areas which are good representations of each of the natural history themes of the park and which will be maintained in a wilderness state. Only certain activities requiring limited primitive visitor facilities appropriate to a wilderness experience will be allowed. Limits will be placed on numbers of users. No motorised access will be permitted. Management actions will ensure that visitors are dispersed.
- *Zone 3: Natural Environment.* Areas that are maintained as natural environments, and which can sustain, with a minimum of impairment, a selected range of low-density outdoor activities with a minimum of related facilities. Non-motorised access will be preferred. Access by public transit will be permitted. Controlled access by private vehicles will be permitted only where it has traditionally been allowed in the past.
- *Zone 4: Outdoor Recreation.* Limited areas that can accommodate a broad range of education, outdoor recreation opportunities and related facilities in ways that respect the natural landscape and that are safe and convenient. Motorised access will be permitted and may be separated from non-motorised access.
- *Zone 5: Park Services.* Towns and visitor centres in certain existing national parks which contain a concentration of visitor services and support facilities as well as park administration functions. Motorised access will be permitted.

Rollins (1993) illustrates, however, that zoning is primarily natural resource based, and does not define the types or levels of recreational opportunities that can occur within such regions of the park. Visitor management within parks in North America is addressed through a number of preformed planning and management frameworks, including the Recreation Opportunity Spectrum (ROS), the Limits of Acceptable Change (LAC), and the Visitor Activity Management Process (VAMP). The former two are American and the latter is Canadian (see Chapter 4 for further discussion on the management of human use in parks).

Only about 2 per cent of Canada's land mass is protected within the national parks system. National parks, though, represent only one of the many systems in Canada developed to safeguard biophysical resources (others include provincial parks, wildlife reserves, environmentally sensitive areas, areas of natural and scientific interest, biosphere reserves, and world heritage sites). However, all told, these protected areas represent between 6 and 7 per cent of the Canadian landscape, but only 2.6 per cent if one excludes protected areas where logging, mining, and hunting are nevertheless allowed (Rollins and Dearden 1993). Furthermore, it has taken some 110 years to fully establish 38 national parks in Canada. Critics of the parks system are quick to point to the fact that Canadians are moving far too slowly in developing the country's network of parks. In 1987, the Brundtland Commission (the World Commission on Environment and Development) set a standard of 12 per cent for all countries in terms of the amount of territory that needed to be established as parks and protected areas. Such a standard is believed to be a level that would ensure a degree of protection for all the earth's major physiographic regions. Canada's commitment to this

---

## CASE STUDY 3.2

### *Last Mountain Lake National Wildlife Area*

This Saskatchewan wildlife area is significant in that 1,013 hectares of it was designated as the first federal bird sanctuary in North America on June 8, 1887. According to the Canadian Wildlife Service (1995), over 280 bird species have been recorded at Last Mountain Lake during peak migration, including 50,000 cranes, 450,000 geese, and several hundred thousand ducks. In addition, the area provides habitat for 9 of Canada's 36 vulnerable, threatened, and endangered birds, including the whooping crane and peregrine falcon. Last Mountain Lake has been designated as a Wetland of International Importance along with approximately 30 other sites in Canada and 700 locations worldwide (as of 1995). Although conservation is the primary purpose within the area, other, more consumptive activities such as hunting and fishing are allowed in adjacent areas under strict regulation.

---

standard came in the form of the Green Plan (Canada Government, 1990). For the purpose of protected areas, some of the goals outlined within the plan include the following: (1) to establish at least five new national parks by 1996; and (2) to negotiate agreements for the remaining 13 parks required to complete the system by the year 2000. (About 18 of the 39 physiographic regions of Canada have yet to contain a park. Canadian park policy requires that at least one park be established within each of the 39 specific regions.)

The expectations are that Canada, as a developed country of the North, has the knowledge, technology, and initiative to be a world leader in the realm of parks and protected areas. The truth of the matter is that Canada has indeed taken a leading role with respect to policy and management of such areas. Comparatively, though, many of the developing countries of the South have been far more conscientious in the development of their parks systems. Panama, Nicaragua, and Costa Rica, for example, have set aside much higher percentages of their national territories as parks and protected areas. Despite Canada's low productivity in the establishment of parks, public opinion is in favour of the creation of more wilderness areas. Reid *et al.* (1995) illustrate that on the basis of a survey of over 1,500 residents of British Columbia, respondents, on average, would be willing to pay between $108 and $130 annually in taxes for a doubling of designated wilderness areas, and between $149 and $156 for a tripling of wilderness areas.

Parks have evolved, globally, to be managed according to the ecological and human conditions of the environments that they inhabit. National parks in Britain (England and Wales), for example, are managed from a very different perspective from those in other Western societies. According to Henderson (1992: 397), conservation in Britain is based on a 'steady state of human intervention designed to maintain a given habitat at a particular successional stage in perpetuity', which he feels is the most unnatural conservation policy possible. British national parks are essentially living, working landscapes, and administratively it is the responsibility of a National Park Authority (which operates within the local government system and whose members are appointed by the government in consultation with the Countryside Commission) to: (1) preserve and enhance park natural beauty; (2) promote access, use, and enjoyment; and (3) ensure proper practices of agriculture, forestry, and economics (Exmoor National Park 1990). The result is that there is a significant amount of socio-economic activity occurring within the 11 national parks of Britain, owing to the fact that virtually none of the land in England or Wales is excluded from human activity (Phillips 1985). Exmoor National Park, the second smallest park in the system, for example,

**Plate 3.1** On Canada's west coast, loggers and environmentalists have fought over some of the world's largest trees. Since being saved these trees have generated much interest among ecotourists in the region

**Plate 3.2** Many wilderness areas have both natural and cultural heritage value. This petroglyph (painting of a moose) in the Clearwater River Provincial Park, Saskatchewan, was painted by aboriginal people many hundreds of years ago

**Plate 3.3** The wilderness character of Clearwater Park, Saskatchewan, Canada. This park is well known for canoeing and hiking

has a resident population of 10,000 and hosts 2.4 million visitors per year, and is almost 80 per cent privately owned (Exmoor National Park 1990). The park plan goes on to explain the role that major developments, economic development, agriculture and forestry, housing, services and facilities, and tourism play in the context of the park, which is consistent with the types of activities that occur in British parks in general (Phillips 1985). Based on this philosophy of use, tourism therefore has a significant role to play within the national parks and is not questioned to the point that it is in other national parks. However, the tourism industry is still expected to operate within guidelines established by the Countryside Commission and the English Tourist Board (Countryside Commission 1990).

The national parks of Britain, as suggested above, are somewhat of an anomaly in that they do not meet the category II criteria of the IUCN (International Union for the Conservation of Nature and National Resources) on national parks. Internationally, then, guidelines have been established in an effort to both control and direct countries in how best to set aside specific lands and water. Phillips (1985) feels that the national parks of Britain would be better classified as protected landscapes, or category V of the IUCN guidelines (see Table 3.1).

## Protected areas: the international scene

A number of international agencies have become involved in the process of aiding individual countries in the process of identifying candidate natural areas. As the need for more protected areas continues to gain momentum globally, such agencies have found it necessary to categorise areas relative to the types of land use and practices of conservation found in different countries. Britain, for example, does not have the type of landscape that would enable the British to establish the magnitude of wilderness reserves that are found in Australia.

**Table 3.1** *Categories for conservation management*

*Category – Scientific Reserve/Strict Nature Reserve*
Areas with some outstanding ecosystem features and/or species of flora and fauna of national scientific importance, representative of particular natural areas, fragile life forms or ecosystems, important biological or geological diversity, or areas of particular importance to the conservation of genetic resources. Concern is for continuance of natural processes and strict control of human interference.

*Category II – National Park*
A relatively large area where one or several ecosystems are not materially altered by human use, the highest competent government authority has taken steps to prevent or control such alteration, and visitors are allowed to enter, under special conditions for inspirational, educative, cultural, and recreative uses.

*Category III – Natural Monument/Natural Landmark*
Area normally contains one or more specific natural features of outstanding national significance which because of uniqueness or rarity should be protected. Ideally little or no sign of human activity.

*Category IV – Nature Conservation Reserve/Managed Nature Reserve/Wildlife Sanctuary*
A variety of areas fall into this category. Although each has as its primary purpose the protection of nature, the production of harvestable renewable resources may play a secondary role in management. Habitat manipulation may be required to provide optimum conditions for species, communities, or features of special interest.

*Category V – Protected Landscape or Seascape*
A broad category embracing a wide variety of semi-natural and cultural landscapes within various nations. In general, two types of areas, those where landscapes possess special aesthetic qualities resulting from human–land interaction and those that are primarily natural areas managed intensively for recreational and tourist uses.

*Category VI – Resource Reserve (Interim Conservation Unit)*
Normally extensive, relatively isolated, and lightly inhabited areas under considerable pressure for colonisation and greater exploration. Often not well understood in natural, land use, or cultural terms. Maintenance of existing conditions to allow for studies of potential uses and their effects as a basis for decisions.

*Category VII – Natural Biotic Area/Anthropological Reserve*
Natural areas where the influence or technology of modern humans has not significantly interfered with or been absorbed by the traditional ways of life of inhabitants. Management is oriented to maintenance of habitat for traditional societies.

*Category VIII – Multiple Use Management Area/Managed Resource Area*
Large areas suitable for production of wood products, water, pasture, wildlife, marine products, and outdoor recreation. May contain nationally unique or exceptional natural features. Planning and management on a sustained-yield basis with protection through zoning or other means for special features or processes.

*Category IX – Biosphere Reserve*
Intended to conserve representative natural areas throughout the world through creation of global and national networks of reserves. Can include representative natural biomes, or communities, species of unique interest, examples of harmonious landscapes resulting from traditional uses, and modified or degraded landscapes capable of restoration to more natural conditions. Biosphere reserves provide benchmarks for monitoring environmental change and areas for science, education, and training.

**Table 3.1** *continued*

*Category X – World Heritage Site*
To protect natural – and also cultural – features considered to be of world heritage quality; examples include outstanding illustrations of the major stages of earth's evolutionary history, habitats where populations of rare or endangered species of plants and animals still survive, and also outstanding archaeological or architectural sites. Stress on maintenance of heritage values for worldwide public enlightenment, and to provide for research and environmental monitoring.

*Category XI – Wetlands of International Importance (Ramsar)*
Marshes, swamps, and other wetlands of value for flood control, nutrient production, wildlife habitat, and related purposes. Management procedures designed to prevent destruction and deterioration through national agreement to an international convention known as Ramsar after the site in Iran where the convention was initially agreed to by a number of founding countries.

*Source*: Nelson (1991)

The IUCN has been especially effective in charting a course for the planning, establishment, and management of protected areas globally. Its categories for conservation management (see Table 3.1) illustrate the variability of protected areas that have been developed internationally, each of which focuses on different aspects of development (the level of tourism infrastructure and use) and preservation. The extent of the work of the IUCN goes well beyond the categorisation of protected areas, though, to include establishing a system of biogeographical provinces of the world; publishing lists and directories of protected areas; publishing conceptual papers on protected areas; publishing the quarterly journal *Parks*; cooperating with United Nations agencies (e.g. UNESCO); holding international meetings, such as the World Conferences on Parks and Protected Areas; and supporting field projects for the establishment and management of protected areas (Eidsvik 1993: 280).

In addition to the IUCN, a series of other international organisations have been activated for the purpose of conservation. The World Heritage Convention of 1972 (UNESCO) provides for the establishment of natural and cultural sites of outstanding universal value. Such sites are woven into the fabric of existing protected areas and, although the system does not impose any new management criteria on existing parks, it does impose an element of symbolism and prestige for countries that maintain such sites. As of 1994, there were some 350 sites worldwide.

A considerable amount of attention has been paid to biosphere reserves as a means by which to overcome some of the problems related to the use of park environments. The biosphere reserve concept grew out of the 1970 UNESCO general meeting which provided the impetus for the development of the Man and the Biosphere programme meetings, which started the following year, and the first reserves started to appear in 1976. The biosphere reserve concept is founded upon three themes: development, conservation, and research. Spatially, these reserves incorporate three distinct zones: (1) a core area, which is minimally disturbed and strictly protected; (2) a buffer, situated around the core and allowing certain types of resource use that do not disturb the core; and (3) a transition zone, which extends outwards into the adjacent territories with no fixed boundary and allowing a full range of human uses. Seven guidelines define the establishment criteria of biosphere reserves. Biospheres work as a linked network of natural areas; they are representations of the 227 biogeographical provinces; they will be examples of special environments (i.e. natural biomes), large in size to ensure effective conservation; they will act as benchmarks for research, education, and training; they will have some form of legal protection; and they will have incorporated within them existing protected areas.

Eidsvik (1983) illustrates that the biosphere is a unique concept relative to other types of protected areas in that it operates under the premise of an inverted pyramid, where the decision-making does not necessarily occur from a centralised federal authority, but rather from the grassroots level with support at various other levels. Eidsvik (p. 230) described this concept as follows:

> The system is designed to support and cherish the participant, operating at his or her own level. . . . Those services that cannot be provided by individuals or their communities then become the responsibility of local agencies. More specialized services come from the provincial governments – and finally, highly specialized residual services are provided by federal agencies.

In practice there are examples of the success of the biosphere reserve concept in the developed and developing worlds. In Canada, for example, the Long Point Biosphere Reserve and region in Ontario operates in partnership with a variety of municipal, provincial, and federal agencies (Francis 1985) including the Canadian Wildlife Service, Transport Canada, the Ontario Ministry of Natural Resources, the private Long Point Company lands, a conservation authority, a regional municipality and other private land holdings, all in an area of approximately 33,000 ha. Similarly, in Mexico, the Sian Ka'an Biosphere Reserve in the Yucatán Peninsula was developed as a result of local efforts to safeguard this 1.3 million acre territory. A non-profit organisation, the Amigos de Sian Ka'an, was established as a result of the initial reserve establishment deliberations, whose responsibility it is to mediate between the private sector and various levels of government. Tourism, owing to the proximity of the reserve to Cancún, has grown significantly in the past twenty years, posing both a threat and an opportunity. Crucial to the viability of the reserve, therefore, is the cooperation between levels of government, the ENGO, and local people, in coming together to solve common dilemmas and improve the quality of their lives. This has prompted the executive director of Amigos de Sian Ka'an to remark that 'If the people who live in the reserve support it, they will take care of it. If not, no amount of guards will stop them' (Norris 1992: 33). Despite the positives of the biosphere programme, there are those who have suggested that it still exists as a top-down approach to conservation, without the commitment to localism that should exist in such areas (Janzen in Chase 1989).

## Private reserves

The growth of ecotourism since the late 1980s has thus generated a great deal of demand for wilderness space. In many regions, however, public land agencies who own and run national parks and other reserves do not have the resources to provide the facilities to run ecotourism programmes effectively (e.g., trails, accommodations, interpretive signs, and so on). In some cases, the threat from illegal operations, such as poaching on public lands, curtails the development of a viable ecotourism industry.

A good example of the creation of space for ecotourism purposes, which has intensified as of late but nevertheless been an option for ecotourism, is the development of private reserves. In some regions like South Africa and Texas there is a legacy of private reserves. In the former case, Dieke (2001) explains that much of the tourism industry was based on consumptive activities, principally hunting by sport hunters, commercial hunters and subsistence hunters. As resources became more scarce, it was the latter group who were refused hunting rights, providing a great deal of tension between those with access and those without, and also the need to control hunting through private parcels of land. In many cases these landscapes remain in the hands of the economically elite, some of whom have incorporated ecotourism programmes in their operations alongside hunting with the purpose of balancing different forms of land use on their properties. (As an aside, Dieke notes that the historical removal of wildlife resource rights from rural peasants has given way to an emerging policy of redistribution of resource rights on the basis of ecotourism. Here, it is argued, decision-making control, and benefits are greatly sought after by the economically marginalised from such a history of injustice.)

Private reserves have been defined by Langholz and Brandon (2001) as lands which are not owned by governmental bodies, are larger than 20 ha in size, and are maintained in a mostly natural state. While this topic appears to be one which is greatly under-researched in the area of ecotourism, there are a few who have provided some important information. For example, in a study conducted in 1992, Alderman wrote that the motivations for the development of private reserves appear to be grouped according to profit, habitat protection and research. She noted that many, as suggested above in the case of South Africa, are in fact hybrids which attempt to be involved in ecotourism, extractive industry like forestry, education and agriculture. About 25 per cent of the private reserves sampled by Alderman were involved in tourism alone.

One of the world's most noteworthy private reserves for ecotourism is the Monteverde Cloud Forest Reserve in Costa Rica. This parcel of land was originally protected during the 1950s by American Quakers who purchased 1,200 ha of land with the purpose of protecting the rich forests from human encroachment. Initially this protection came about to maintain sufficient lands for the dairy operation undertaken by the tenants, but later expanded with a conservation mandate given pressure by biologists and other conservationists who recognised the biological value of the land. Honey (1999) notes that tourism was initially quite low key, especially up to the 1970s: only 471 visitors in 1974, climbing to 3,257 in 1980, 11,762 in 1985, and later to 49,580 in 1992, coinciding with the growth of ecotourism throughout the 1980s and early 1990s. Through land purchases financed by various conservation organisations, the reserve has grown to 10,522 ha encompassing eight ecological zones.

## CASE STUDY 3.3

### Urban ecotourism

In one of the classic early papers on tourism, Christaller (1963) wrote that tourism by nature avoids central places in favour of the natural resource base found outside the limits of the city. This, he reasons, is attributable to the notion that cities have become urbanised and overpopulated to the point that quality of living could be better realised through such excursions. Peripheral regions were thus seen by Christaller as the antithesis of cities. By its nature, tourism has facilitated the process for capitalism to achieve further growth in new sectors and in new places. As suggested by Husbands 'Tourism is . . . an example of such a new activity [in the periphery] and the peripheral space provided by it is an example of the production of space, and the further penetration of the periphery by capitalism' (Husbands 1981: 51). This new space has been most notably used for sun, sea and sand tourism (e.g., Caribbean), but also more recently in regions which support adventure travel and ecotourism.

Given the foregoing discussion, it is interesting to note an emerging focus on ecotourism in urban regions. This phenomenon may be defined as

*continued*

travel and exploration within and around an urban area that offers visitors enjoyment and appreciation of the city's natural areas and cultural resources, while inspiring physically active, intellectually stimulating and socially interactive experiences; promotes the city's long-term ecological health by promoting walking, cycling, public transportation; promotes sustainable local economic and community development and vitality; celebrates local heritage and the arts; is accessible and equitable to all.

(Blackstone Corporation 1996: 2–5)

Such an elaborate definition, casting as wide a net as it does, underscores the importance of nature, culture, physical activity, sustainability and community development in urban areas. Some of the most logical places for urban ecotourism include parks, cemeteries, golf courses, sewage lagoons and stormwater control ponds, landfill and waste disposal sites, high rise and other structures, and zoos and botanical gardens (Lawton and Weaver 2001). Good examples of these include El Avila National Park in Caracas, Venezuela, which lies just to the north of the city, containing hundreds of well-kept trails and providing excellent views of the coast and of Caracas; Century City in South Africa, and New York's Central Park.

Perhaps one of the most intriguing issues concerning this phenomenon relates to the philosophy which ecotourism attempts to uphold, and the urban environment which, as stated above, has represented just about everything which ecotourism is not. In approaching the issue with an open mind, it is nevertheless important to examine the underbelly of examples of urban ecotourism in deciding if they are indeed representative of ecotourism's ecocentric ideals. As stated by Lawton and Weaver (2001), golf courses (and cemeteries for that matter) are notorious for their use of pesticides in keeping lawns green. Although these regions may harbour keynote species like eagles, alligators, and bears, the volume of chemicals introduced to the environment raises serious questions about the ecological health of the entire ecosystem. This can also be the case with the concept of the urban marineland habitat, which contains a variety of marine mammals like killer whales. Unfortunately, however, these animals are unable to live out their life cycles fully because of a variety of issues related to captivity. Conversely, many authors (Lawton and Weaver 2001; Higham and Lück 2002) have identified a number of innovative developments in urban environments which have stimulated ecotourism through the creation of space, much in the way described by Husbands earlier. Municipalities have reclaimed mines as well as waste disposal sites and turned these into green belts, including parks and golf courses. These have become important sites for native flora and fauna, and have quite often contributed to developments which have allowed managers to link green belts (corridors) throughout the urban environment. Higham and Lück (2002) have taken up this issue in suggesting that many groups have made significant contributions to conservation through the restoration of natural areas which have been degraded by human intervention, making ecotourism in the urban environment an option.

As a case in point, Canada's Green Tourism Association (Toronto) argues that the principles of ecotourism applied to urban tourism may actually be better for the environment in certain places because cities are able to absorb the impacts of tourism when compared to wilderness areas (Dodds and Joppe 2001). The city of Toronto has approximately 20,000 acres of green space within the city, including about 370 species of birds. Examples of ecotourism include the Leslie Street Spit, which acts as a migratory flyway; R.C. Harris Filtration plant; the Humber Rivers; self-guided discovery walks throughout the city; the Don River restoration project; and a number of bike, blade and heritage walks provided by the conservation authority. In addition, the Green Tourism Association, in concert with the city, published the '"Other" Map of Toronto', which is a green map linking tourism with the natural environment. In February of 2002, the Green Tourism Association participated in the International Outdoor Adventure Show in Chicago. Here they conducted a study which revealed that 71 per cent of visitors said they would use businesses, restaurants and accommodation that made provisions to minimise negative environmental and social impacts. In a previous study carried out in July 2000, it reported that 91 per cent of respondents said that when they visit a city they 'sometimes or often' visit green spaces and park lands. The study also observes that 83 per cent of those polled felt that ecotourism could take place in a city (Green Tourism Association 1999).

It remains to be seen how effective cities will be in the development of ecotourism. There is little question that the greening of cities is an urban response to sustainable development, and more specifically sustainable tourism. Perhaps one of the main constraints for ecotourism in the urban environment is overcoming the idea that tourism, from a spatial context, is a place for business and culture, and less for adventure and ecotourism. There is also the argument, although controversial, that tourist dollars might better be spent in peripheral areas, possibly helping to stimulate the economies of marginalised peoples – acknowledging that there are many intervening factors which might catalyse or constrain people from venturing into more peripheral areas. However, if planners are able to integrate key aspects of conservation, environmental education, community welfare (broadly defined), ethics, and sustainability, then some might have cause to suggest that there are few reasons why the urban realm might not be an acceptable venue for ecotourism.

## Ecosystem management and protected areas

As identified previously in Figure 3.1, parks management has evolved significantly over time. One of the finest examples of this evolution is the development of the ecosystem management philosophy, which has blossomed as a consequence of the realisation that in order to effectively safeguard an environment one must scientifically understand the relationships and processes that exist within such a setting. The biodiversity crisis, new ecological theories, and dissatisfaction with governmental regulatory measures also contributed to the birth of this mode of thinking (Grumbine 1996). Foremost, biological and social systems theory became the foundation of ecosystem management, once it became clear that ecological sustainability can be attained only through substantial societal change. Indeed, as Francis (n.d.) suggests, we must begin to see ourselves as integral components of complex ecosystems, components in turn related to one another over a variety of spatial and temporal scales. The feeling is that the species with which we share ecosystems have inherent value in and of themselves, and should not be judged on the basis of their ability to provide us with resources, but rather as important elements within the complex system.

In definition, ecosystem management, according to the Canadian Environmental Advisory Council (CEAC), as it relates to its important function in the planning and management of parks and protected areas, refers to the 'integrated management of natural landscapes, ecological processes, wildlife species and human activities, both within and adjacent to protected areas' (CEAC 1991: 38). The following definition, put forward by Johnson and Agee (1988), emphasises the inclusion of both social and ecological processes that help shape and transform ecosystems. Hence, we can never separate human and biophysical elements within an ecosystem; they are inseparable. As a species, we live in or visit just about every inhabitable space on the planet that we can.

Ecosystem management involves regulating internal ecosystem structure and function, plus inputs and outputs, to achieve socially desirable conditions. It includes, within a chosen and not always static geographic setting, the usual array of planning and management activities but conceptualised in a systems framework; identification of issues through research, public involvement, and political analysis; goal setting; plan development; use allocation; activity development (resources management, interpretation); monitoring; and evaluation. Interagency coordination is often a key element of successful ecosystem management, but is not an end in itself. Success in ecosystem management is ultimately measured by the goals achieved, not by the amount of coordination.

(Johnson and Agee 1988: 7)

According to Chipeniuk (1988), the management of parks from an ecological perspective is not complete, and ecosystems are not natural, because they lack a principal component of these ecosystems that existed at the time of European contact: human hunters and gatherers who played a role as predators. Chipeniuk argues that this is a niche which, for biological and socio-cultural reasons, must be filled by human surrogates. In a biological sense, the environment would be returned to more of a natural state (i.e. pre-contact); while from the socio-cultural context the return of people as part of the functioning of park ecosystems would cease to delude people about the proper place of human beings in the natural world. This point is touched on by Pretty and Pimbert (1995: D8), who argue that when local indigenous people, for example, are excluded from conservation the goals of conservation are at risk. They write that 'some "pristine rain forests," assumed to be untouched by human hands, are now known to have once supported thriving agricultural communities. The "pristine" concept of the wilderness is an urban myth that exists only in our imagination.'

The study of the relations between humans and their respective environments, or human ecology (Burch 1988), has been described as a field of research that has potential to institute the human dimensions in ecosystem management along with the biophysical. In fact Nelson (1993) describes ecosystem management and human ecology as different sides of the same coin – where human ecologists and biologists should find common ground. Nelson (ibid.: 74) has identified five ways in which human ecology can aid in ecosystem management:

1 by presenting a historical understanding of an area in terms of nature and humans and their interactions;
2 by doing a history systematically in terms of the culture which defines humans, e.g. policies and institutions, perceptions, attitudes, values, technology;
3 by presenting the history spatially in terms of similarities and differences over space;
4 by linking human studies to concepts or ideas that are the concern of other professionals, for example the concept of landscape, which has roots in architecture, geography, geology and other fields such as biology; and
5 by presenting historical understanding in terms that are meaningful and attractive to a wide range of citizens; by drawing people to the human–nature interface – to the dynamics of ecosystem management, from a human perspective – to complement those people drawn to it from a biological or scientific perspective.

Peterson (1996: 27, 28) takes a more hard-hitting approach to linking human ecology and ecosystem management. He advocates four extreme approaches to human ecology that humans use to define their relationship with the planet. These include:

1 *Dominion*. This implies rule by a monarch, where humans are held to be in charge of the earth. They can be exploitative or serve the well-being of all, in their role as 'king'.

2   *Stewardship*. This paradigm puts humans in the role of caretaker of the earth, managing the earth in trust as an agent for some employer or client. The client may be the human race or the earth itself.

3   *Participation*. This role sees humans in symbiosis with other species of the planet, so that a position of equity is conveyed. Humans serve by constructing cooperative and complementary relationships through which all other species are better off.

4   *Abdication*. Here, all rights to prosper are relinquished when such rights conflict with the functional values of other species. Humans are caught in the predator–prey relationship just as other species are.

The implications of these four approaches to ecosystem management, according to Peterson, are that effective human management of ecosystems will not occur without a clear understanding of the place of people in the context of the ecosystem.

A good example of the application of human ecology principles to parks management is found in the work of Slocombe and Nelson (1992). These authors identify a number of pressures that exist within parks and protected areas as a result of internal (illegal entry, removal of flora and fauna) and external (pollution, mining) threats. Their paper examines three protected areas (Kluane National Park in Canada; Wrangell–St Elias National Park and Preserve in the United States; and Kakadu National Park in Australia) in developing management responses to issues within these areas. Eight management criteria termed dimensions of difference, were developed to assess and compare the parks, namely access, tourism, resource extraction, aboriginals' role, administration issues, scientific research, interpretation, and regional integration. Slocombe and Nelson conclude that interest is growing in human ecological approaches to parks and protected areas, as they apply to many regions around the world.

Although human ecology is a well-established discipline, its place in tourism studies has not been well represented. For example, in a recent review of the journal *Human Ecology*, I found just two examples of human ecology and tourism. Both papers explored the resource competition that exists between recreational anglers and local users of the resource base. Berkes (1984) discovered that although no ecological competition existed between local and non-local people, a type of perceptual competition existed between the two. Conversely, de Castro and Bergossi (1996) found that competition existed between tourists and locals over certain fish species, but only at certain times of the year. In an effort to better link tourism studies and human ecology, Fennell and Butler (in press) have recently developed a human ecological approach to tourism group interactions by suggesting that various forms of tourism (e.g. ecotourism, general interest tourism) share specific relationships (predatory, competitory, neutral, and symbiotic) with each other, local people, other land users, and the tourism industry, all within the context of a network of land, including parks and protected areas, town sites, and other settings. The paper explores the notion that each of the various stakeholders identify in a tourism setting value and use resources differently and therefore place different levels of pressure on each other and the resource base.

## Conclusion

The philosophical issues related to the place and role of humans in the environment have been explored most effectively through conservation and preservation. Parks as manifestations of the development of conservation must continue to act as test sites for human–environment interactions. We must continue to employ new strategies to enable people to strike a fair and equitable balance between use and preservation in a world that perpetuates the value of human beings at the expense of other life forms, and which continually encroaches upon the earth's most sensitive and significant regions, through tourism and other land uses. The principles put forward in ecosystem management and human ecology may help us to better understand the place of humans in the natural world. However, this technical and scientific information must be analysed in the context of appropriate philosophical and conceptual questions which we have been asking, as evident in the writings of many of the Romantic scholars, for some time.

## Summary questions

1   What is a natural resource, and why are natural resources so important to the ecotourism industry?
2   Discuss the differences which exist between conservation perspectives based on efficient use, spirituality, and harmony. Name some historical figures who were representative of these positions.
3   Discuss how park management has evolved over time. What are the differences between internal and external park threats, and what are the implications to park management?
4   Why are park zones so important, and what are their implications for the human use of parks and protected areas?
5   What is the difference between public reserves and private reserves? Why are private reserves a better alternative for ecotourism in some countries?

# 4 The social and ecological impacts of tourism

Tourism research has typically centred around topics related to the social, ecological, and economic impacts of the tourism industry. Social impact studies have usually involved an analysis of how the industry has affected local people and their lifestyles, whereas ecological studies have tended to emphasise how the industry has transformed the physical nature of local and regional landscapes. Such studies seem to be in contrast to tourism economic research, which in most cases tends to illustrate the income-generating power of the industry within the community, region or country. Given that this impact research is quite voluminous, it is not the purpose of the following discussion to provide a complete overview of research in these areas. Instead, this chapter will focus most extensively on issues related to ecological impacts, carrying capacity and, less specifically, on social impacts. Economics and marketing in ecotourism are the topic of Chapter 5.

## Social impacts of tourism

One of the most persuasive socially oriented frameworks developed to analyse the impact that tourism has on local people and their environments is based on the work of Doxey (1975), who, in a general context, was able to encapsulate the evolving sentiment that local people express as tourism expands and occupies greater proportions of a local economy over time. Doxey wrote that there are essentially four main stages to consider in the assessment of local feelings toward the tourism industry. These include:

1   *Euphoria*. Tourists are welcomed, with little control or planning.
2   *Apathy*. Tourists are taken for granted, with the relationship between both groups becoming more formal or commercialised. Planning is concerned mostly with the marketing of the tourism product.
3   *Annoyance*. As saturation in the industry is experienced, local people have misgivings about the place of tourism. Planners increase infrastructure rather than limit growth.
4   *Antagonism*. Irritations are openly displayed towards tourists and tourism. Planning is remedial, yet promotion is increased to offset the deteriorating reputation of the destination.

There are myriad examples of regions that have been subject to this form of cycle within tourism (see also Butler, 1980, later in this chapter). As a case in point, Bermuda experienced visitor numbers of some ten times its local population in 1980 (some 600,000 people) in an area approximately 21 square miles in size. This type of tourist-to-local ratio is indicative of the conditions that have led to social conflict (as identified by Doxey). Although such a proliferation of visitation no doubt has its economic rewards, what the host country gives up to attract tourism dollars cannot be measured simply in economic terms. It is no accident that the most vital and creative parts of the Caribbean, for example, have been precisely those that have been most touched by tourism (Chodos 1977: 174). The oft-quoted claim of Evan Hyde, a Black Power leader in Belize in the early 1970s, that 'Tourism is whorism' (Erisman 1983: 339) reflects the frequent claims that tourism leads to conflict between locals and hosts.

A notable impact of tourism on traditional values is the demonstration effect (Britton 1977; Hope 1980; Mathieson and Wall 1982), where local patterns of consumption change to imitate those of the tourists, even though local people only get to see a side of tourists that is often not representative of their values displayed at home (e.g. spending patterns). Alien commodities are rarely desired prior to their introduction into host communities and, for most residents of destination areas in the developing world, such commodities remain tantalisingly beyond reach (Rivers 1973). The process of commercialisation and commodification may ultimately erode the local goodwill and authenticity of products, as identified by Britton (1977):

> Cultural expressions are bastardized in order to be more comprehensible and therefore saleable to mass tourism. As folk art becomes dilute, local interest in it declines. Tourists' preconceptions are satisfied when steel bands obligingly perform Tony Orlando tunes (and every other day the folklore show is narrated in German).
>
> (Britton 1977: 272)

Such a fragmentation of culture has been found to occur on many levels within destinations, most notably from the standpoint of prostitution; crime; the erosion of language in favour of more international dialects; the erosion of traditions, either forgotten or modified for tourists; changes to local music and other art forms; food, in the form of a more international cuisine; architecture; dress; family relationships (e.g. young children earning more than their parents from toting bags at airports); and, in some cases, religion. In recognising the potential for social impact within a tourist region, Ryan (1991: 164) has identified a number of key points, all of which may be used as indicators or determinants of impact. These are as follows:

1 the number of tourists;
2 the type of tourists;
3 the stage of tourist development;
4 the differential in economic development between tourist-generating and tourist-receiving zones;
5 the difference in cultural norms between tourist-generating and tourist-receiving zones;
6 the physical size of the area, which affects the densities of the tourist population;
7 the extent to which tourism is serviced by an immigrant worker population;
8 the degree to which incoming tourists purchase properties;
9 the degree to which local people retain ownership of properties and tourist facilities;
10 the attitudes of governmental bodies;
11 the beliefs of host communities, and the strengths of those beliefs;
12 the degree of exposure to other forces of technological, social, and economic change;
13 the policies adopted with respect to tourist dispersal;
14 the marketing of the tourist destination and the images that are created of that destination;
15 the homogeneity of the host society;
16 the accessibility to the tourist destination; and
17 the original strength of artistic and folkloric practices, and the nature of those traditions.

As ecotourism continues to diversify and exploit relatively untouched regions and cultures, there is the danger that a cycle of events similar to that identified by Doxey will occur. The lessons from the Caribbean model of tourism development, for example, are that the industry must tread lightly in securing an equitable relationship between how the industry is planned and developed and the needs of local people. Britton (1977) recognised the importance of small-scale, local architecture, tourism zoning, gradual growth, reliance on locally produced goods, joint ventures, and a diversification in the market, in releasing the Caribbean from metropolitan domination (see Chapter 5). All these elements, as identified in Chapter 1, are indicative of the alternative tourism development paradigm to which ecotourism must subscribe.

# Ecological impacts

## The early years

Concern over the ecological effects of tourism started to mount during the 1960s and 1970s (Pearce 1985), through the realisation that the industry had the capability of either moderately altering or completely transforming destination regions in adverse ways. For example, the *National Geographic Magazine* as far back as the early 1960s (Cerruti 1964) was enquiring as to whether Acapulco had been spoiled by overdevelopment, while Naylon (1967) discussed the need to alleviate some of the stress caused by a high concentration of tourism in the Balearic Islands and the Costa Brava in Spain by employing regional development strategies designed to promote other areas that were as yet undeveloped. Pollock wrote that although tourism had begun to play an important role in the economy of Tanzania, the 'vital necessity for game conservation in the interests of ecology, tourism, game farming and ranching, and for moral, aesthetic, philosophical and other reasons has been recognized increasingly both at national and international levels' (Pollock 1971: 147). Others have commented on the physical impacts of tourism in city and regional environments, including Harrington (1971), who illustrated that the unregulated development of hotels in London threatened the quality of life in the city, and Jones (1972), who makes reference to tourism development as a classic case of the battle that exists between conservation and preservation on the island of Gozo. Crittendon (1975) illustrates that while tourism has transformed much of the world's natural beauty into gold, the industry may have planted the seeds of its own destruction.

Sensitivity to environmental issues in the realm of tourism studies gained a tremendous boost in the mid-1970s from the efforts of Budowski (1976), Krippendorf (1977) and Cohen (1978) in their work on tourism and the environment. Budowski identified three different 'states' in tourism's relationship with environmental conservation: conflict, coexistence and symbiosis. He felt that tourism's expansion resulted in an unavoidable effect on the resources upon which it relied, and therefore felt that the relationship at the time was one of coexistence moving towards conflict. Krippendorf was one of the first to write on the importance of planning, and the dispersion of tourists and tourism developments, as a means by which to minimise impacts, while Cohen reviewed the work to date (academic and non-academic) on tourism and the environment. He speculated on the apparent 'mood of the day' by insisting that there was indeed a distinct difference between development for purposes of improvement and aesthetic appeal versus the vulgar, undesirable, and irreparable damage created by modern tourism.

More research on the ecological impacts of tourism began to emerge in the early 1980s from Krippendorf (1982), who, like Budowski, recognised that the resource base acted as the raw material of tourism, which through improper use and overuse loses its value.

Krippendorf cited ski-slopes, holiday villages, camping and caravan sites, and airfields as examples of developments that when fully functional seem to subsume the environment forever for their own uses. Travis (1982) suggested in his review of literature that while most studies on tourism concentrated on the economic benefits of tourism, there was also a tremendous range of topics related to its negative impact, including pollution, crowding and congestion, damage/destruction of heritage resources, land use loss, ecosystem effects, loss of flora and fauna and increased urbanisation. Concurrently, Coppock (1982) identified similar areas in which tourism has had an adverse impact on nature conservation in the UK. These were identified as loss of habitat, damage to soil and vegetation, fire, pollution, and disturbance of flora and fauna. In the 1980s books started to emerge that dealt with the development and impacts of tourism, including Pearce's *Tourist Development* (1991) and Mathieson and Wall's work on economic, social, and ecological impacts (1982).

Tourism research on ecological impacts further intensified throughout the 1980s on the basis of a wealth of information surfacing on the relationship between tourism and conservation, and the need to address how best to overcome tourism's negative impacts. Romeril (1985) wrote, in a special edition on tourism in the *International Journal of Environmental Studies*, that concern for the environmental impacts of tourism has come on the wings of a broader global concern over the conservation of natural resources generated by the United Nations Human Environment Conference of 1972, the World Conservation Strategy of 1980, the Report of the Brandt Commission (1980), and the Manila Declaration on World Tourism in 1980, which stated that:

> The use of tourism resources could not be left uncontrolled without running the risk of their deterioration, or even destruction. The satisfaction of tourism requirements must not be prejudicial to the social and economic interests of the population in tourist areas, to the environment and above all to natural resources which are the fundamental attractions of tourism and historical and cultural sites. All tourism resources are part of the heritage of mankind.
>
> (cited in Romeril 1985: 216)

In the same edition, Pearce (1985) reproduced a framework for the study of environmental stress that was established by the OECD in 1981, and included stressor activities, the pressure resulting from the activity, the primary environmental response, and the secondary human response or reaction of the stress. Four main examples were identified in this framework related to permanent environmental restructuring, generation of waste, tourist activities, and effects on population dynamics, as shown in Table 4.1.

One of the most complete overviews of the history of ecological concern in the tourism industry was written by Shackleford (1985). His review of tourism and the environment suggests that the International Union of Official Travel Organisations, or IUOTO (the precursor to the WTO), had been working with the environment in mind since the early 1950s, through the efforts of the Commission for Travel Development. From 1954 onwards, the protection of heritage was an agenda item for this organisation. Subsequent work by IUOTO led to the recommendation by its Fifteenth General Assembly that world governments implement the following 1960 resolution:

> The General Assembly, considering that nature in its most noble and unchanging aspects constitutes and will continue increasingly in the future to constitute one of the essential elements of the national or world tourist heritage. Believes that the time has come for it to deal with the problems raised by the dangers threatening certain aspects of nature. . . . Decides consequently to recommend to all IUOTO Member Countries

**Table 4.1** *A framework for the study of tourism and environmental stress*

| Stressor activities | Stress | Primary response: environmental | Secondary response: (reaction) human |
|---|---|---|---|
| 1 *Permanent environmental restructuring*<br>(a) Major construction activity<br>• urban expansion<br>• transport network<br>• tourist facilities<br>• marinas, ski-lifts, sea walls<br><br>(b) Change in land use<br>• Expansion of recreational lands | Restructuring of local environments<br>• expansion of built environments<br>• land taken out of primary production | Change in habitat<br>Change in population of biological species<br>Change in health and welfare of man<br>Change in visual quality | *Individual* – impact on aesthetic values<br>*Collective measures*<br>• expenditure on environmental improvements<br>• expenditure on management of conservation<br>• designation of wildlife conservation and national parks<br>• controls on access to recreational lands |
| 2 *Generation of waste residuals*<br>• urbanisation<br>• transportation | Pollution loadings<br>• emissions<br>• effluent discharges<br>• solid waste disposal<br>• noise (traffic, aircraft) | Change in quality of environmental media<br>• air<br>• water<br>• soil<br>Health of biological organisms<br>Health of humans | *Individual defensive measures* local residents<br>• air conditioning<br>• recycling of waste materials<br>• protests and attitude change towards tourists<br>• change of attitude towards the environment<br>• decline in tourist revenues<br>*Collective defensive measures*<br>• expenditure on pollution abandonment by tourist related industries<br>• clean-up of rivers, beaches |
| 3 *Tourist activities*<br>• skiing<br>• walking<br>• hunting<br>• trail bike riding<br>• collecting | Trampling of vegetation and soils<br>Disturbance and destruction of species | Change in habitat<br>Change in population of biological species | *Collective defensive measures*<br>• expenditure on management of conservation<br>• designation of wildlife conservation and national parks<br>• controls on access to recreational lands |
| 4 *Effect on population dynamics*<br>Population growth | Population density (seasonal) | Congestion<br>Demand for natural resources<br>• land and water<br>• energy | *Individual*<br>• Attitudes to overcrowding and the environment<br>*Collective*<br>• Growth in support services, e.g. water supply, electricity |

*Source*: Pearce (1985)

to exercise increased vigilance regarding the attacks made on their natural tourist resources.

(Shackleford 1985: 260)

Other examples of environmental impact research in tourism in the 1980s include work by Farrell and McLellan (1987) and Inskeep (1987) in a special edition of the *Annals of Tourism Research*. Their research suggests that planning and policy are critical components of a more ecologically based tourism development strategy for the future (more on policy in Chapter 6). For example, Inskeep (1987) writes that determining the carrying capacity of tourist sites is an important factor in the planning and design of appropriate tourist facilities, a concept around which Mlinarić (1985) built a discussion on tourism and the Mediterranean.

Up to and including the 1980s, few models had attempted to study tourism impacts from an ecological standpoint. This notion is reinforced by Getz (1986), who identified only three ecologically based frameworks in an analysis of over forty tourism models. These included a comprehensive model by Wall and Wright (1977), the OECD model mentioned above, and a unique model by Murphy (1983), who made an analogy between the tourism industry (locals, the industry and tourists) and predators and prey interacting within an ecosystem. Although Getz's work was completed some years ago, Dowling (1993) reports that little had changed with respect to the implementation or creation of tourism development models from the environmental disciplines. Fennell and Butler (in press) point to the fact that because it is largely social scientists making inferences on ecological matters, there is much uncertainty with respect to the ecological impacts of tourism. They also point to the fact that there is virtually no natural science research emerging from the tourism journals to aid in the continuing struggle to come to grips with the tourism impact dilemma, with the result being that impacts are often anticipated but not controlled (see also McKercher 1993b).

An excellent addition to the literature on the environmental impacts from tourism has come in the form of a book by Newsome *et al*. (2002), who identify a whole range of different types of impacts, their sources, and regions in which these take place. The authors note that sources of impacts cited in the literature include trampling (vegetation, microbes, soils) access roads and trails, built facilities and camp grounds (camp sites and firewood), water edges (river banks, lakes and reservoirs, coastal areas and coral reefs). The book recognises the importance of looking at impacts from a geographical perspective, by identifying specific ecoregions, including mountains, caves, arctic–alpine environments, tropical realms, and arid environments. Others, such as Tribe *et al*. (2000), have identified a whole range of recreational activities and their associated positive and negative impacts along the following lines: habitat change/loss, species change/loss, aesthetics, physical pollution, soil change/ damage, noise pollution, conflicts, energy/water usage, local community, and revenue vs. costs. It is thus important to realise the inferences that go along with a discussion of positive and negative environmental impacts of tourism. While most research examines impacts from the negative context, and justifiably so, policy-makers and academics need to be aware of the fact that impacts occur along a continuum, and that these are not discrete occurrences but rather those which may be determinable along social and ecological lines, as suggested in the next section.

# The concept of carrying capacity

Increasingly, researchers and practitioners have begun to recognise the dangers inherent in accommodating an increasing number and diversity of experiences for a growing consumer-based society. It is in an agency's best interests to be aware of and sensitive to the broad range of different user groups (non-recreational, and recreational including consumptive and non-consumptive) in a setting and their various needs. Over time, managers have begun to learn that sound planning and development of public and private lands must be viewed as the best means by which to ensure the safety of the resource base first, even over the needs and expectations of participants. These types of issues have been raised and debated extensively through the literature on carrying capacity.

The concept of carrying capacity is not new. Butler *et al.* (1992) argue that for some time people have worried about their excessive use upon stocks of game and other renewable resources, as suggested by this sixteenth-century poem:

> But now the sport is marred,
> And wot ye why?
> Fishes decrease,
> For fishers multiply.

In the strictest ecological sense, species maintain a balance between birth and death, and predator–prey relationships within an ecosystem. It is the human factor and the manipulation and exploitation of resources that offset this balance. Generally speaking, the concept of carrying capacity can be loosely defined on the basis of the following four interrelated elements: (1) the amount of use of a given kind (2) a particular environment can endure (3) over time (4) without degradation of its suitability for that use.

In the early 1960s the concept was applied recreationally for the purpose of determining ecological disturbance from use (Lucas 1964; Wagar 1964). However, it was quickly discovered that an understanding of ecological impact might be achieved only through the consideration of human values, as evident in the following passage:

> The study . . . was initiated with the view that the carrying capacity of recreation lands could be determined primarily in terms of ecology and the deterioration of the areas. However, it soon became obvious that the resource-oriented point of view must be augmented by consideration of human values.
>
> (Wagar 1964: 23)

This prompted researchers to try to balance the importance of both environmental impacts and human perceptions in their various interpretations of the carrying capacity concept.

Typically, environmental impacts can be objectively measured through an analysis of ecological conditions. In the outdoor recreation literature, a value judgement has been placed on the term 'impact', denoting undesirable change in environmental conditions (Hammitt and Cole 1987). Concern lies in understanding the type, amount, and rate of impact on the resource base through recreational use. A campsite, for example, may be severely impacted over time by accommodating high levels of use. Significant changes may occur to the ecology of the site as evident through the compaction of soil (e.g. exposing roots and increasing erosion), vegetation (e.g. using both dead and live tree limbs for the construction of fires, and trampling saplings), wildlife (e.g. habitat modification, and animal harassment), and water (e.g. the addition of human waste and chemical toxins to the aquatic environment). The heaviest impact to a campsite, however, occurs during the first couple of years of use, and impact subsides over time as the site becomes harder and harder (see Figure 4.1). These data provide strong evidence to suggest that new campsites ought not to be developed, but rather that existing ones (i.e. directing use to these areas) ensures the least amount of disruption to the resource base.

From the sociological perspective, carrying capacity becomes much more dynamic and difficult to measure. The complications arise when considering the level or limit to

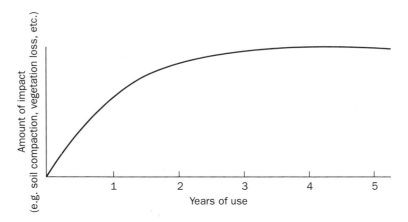

**Figure 4.1** Impact on recreation sites

*Source*: Hammitt and Cole (1987)

the amount of use which is appropriate for a specific resource. Owing to the nature of the resource as a subjective, perceptual entity, different types of users will have different needs and expectations of the resource. Consequently, the tolerance of these user groups (e.g. jet-boaters and canoeists) to one another will vary. To compound the matter further, the tolerance of individuals within groups (intragroup tolerance) will also vary. To take canoeists as an example, each individual within this recreational group will also have certain experiential expectations. Encounters with other canoe parties (or other user types), the density of use (the number of users per unit area), and the perception of crowding (the behavioural response to such encounters) will differ for these individuals over space and time.

Researchers and managers have argued consistently that the goal of recreation management is to maximise user satisfaction (Lucas and Stankey 1974). Despite this agreement, past research has generally failed to document the empirical relationships between use levels and visitor satisfaction deemed necessary for the development of evaluative standards for the management of a resource. Shelby and Heberlein (1986) measured perceived crowding and satisfaction through the importance of use levels and encounters in their analysis of river rafters, canoeists, tubers (people who float down rivers on rubber tubes), fishers, deer hunters, and goose hunters in the western United States. Use levels provided an objective measurement that evaluated how many people were using the resource. Encounters were determined by having a researcher follow groups and count the number of contacts they had with others, or by simply asking users to report contacts with others. The authors hypothesised that:

1   As use levels and encounters increased, perceived crowding would increase.
2   As use levels and encounters increased, satisfaction would decrease.

Their findings illustrate that higher use levels (the number of people using a resource) do not always make people feel more crowded. There was a stronger relationship between contacts and perceived crowding. Generally, people felt more crowded as contacts increased for all activities except rafting when compared with use levels. This is expected because the number of people one actually sees should have a greater impact than the overall number using the area. Crowding means too many people, but use levels and contacts do not entirely explain feelings of crowdedness.

**Plate 4.1** Wilderness users are wise to use existing campsites rather than create their own in relatively untouched park regions

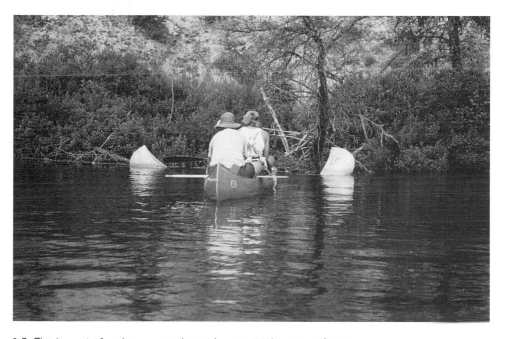

**Plate 4.2** The impact of park users on the environment takes many forms

**Plates 4.3 and 4.4** Those who maintain a strict definition of ecotourism would suggest leaving plants and animals alone altogether, which means a refusal to pick them up

The authors also used the use level and encounter variables to test whether or not satisfaction decreased with increasing levels of use. Results suggest that recreationists were just as satisfied at high use levels as they were at low use levels. In fact, in all cases low-use-level visitors were not significantly more satisfied than high-use-level visitors. A number of authors, including Shelby and Heberlein (1986), Pitt and Zube (1987), Stankey and McCool (1984) and Graefe *et al.* (1984a), indicate that the weak relationship between satisfaction and perceived crowding occurs for a number of differing normative/perceptual reasons. They offer the following as explanations for this poor relationship:

- *Self-selection.* People choose recreational activities they enjoy and avoid those they do not. There is an expected high level of satisfaction, regardless of the use level, because people will select experiences they will enjoy.
- *Product shift.* Users may change their definitions of recreation experiences to cope with excessive encounter levels. As a result, they may remain satisfied as contacts increase. In addition, the contacts themselves may play a role in changing the definition of the situation (hikers seeing more people in the wilderness may change the definition of the experience).
- *Displacement.* Individuals who are truly sensitive to high-density relationships may have already moved out of the environment being studied to a less intensively used area, being replaced by those less sensitive to high density.
- *Multiple sources of satisfaction.* Satisfaction is a broad psychological construct. The number of other people is only one of many things that might affect satisfaction or dissatisfaction.
- *Rationalising.* Recreationists may make the best of even a bad situation, focusing on positive aspects and minimising those that are less pleasant. People who complain about the number of others on a river are still likely to have a good time by learning to ignore the negative aspects of seeing others.
- *Activity-specific influences.* The response of individuals to contacts with others may vary according to the types of activities and behaviour encountered. An individual may be quite tolerant of contacts with hikers and extremely intolerant of contacts with off-road vehicles. The extent to which one type of use impacts another depends upon the social and personal norms visitors use to evaluate the appropriateness of specific behaviours.
- *Conceptualisation and measurement of satisfaction may be inadequate.* The multi-dimensional character of experience, by definition, makes the likelihood of high correlations between a unidimensional overall satisfaction scale highly unlikely. Research is beginning to show that people can be satisfied and dissatisfied with their experience at the same time. Graefe *et al.* (1984b) found that 71 per cent of visitors to a Recreation Wilderness Area considered their trip excellent or perfect. However, 41 per cent also included the comment that they experienced at least one dissatisfying incident during their visit.

The above seven variables illustrate that the measurement of an individual's level of perceived crowding/satisfaction is difficult to attain. Recreationists may either adjust to a dissatisfying situation through a product shift, adapt to the situation, rationalise the experience, or displace entirely from the site. The social and personal normative values that an individual might use to evaluate a site are unique and specific. This, coupled with inadequate measures of user satisfaction, may create a tremendous void between what managers feel they know about human–resource relationships and what they do not.

Defining and operationalising carrying capacity is further complicated by the necessity of considering management objectives, the effects of use on environmental quality, and the effects of use on user and host desires and expectations (Hovinen 1981; Wall 1982; Stankey and McCool 1984; Haywood 1986; O'Reilly 1986). The findings of Butler *et al.* (1992), in an extensive review of literature, concur that the concept of carrying capacity requires adept management. No mythical figure exists for limiting the amount of use in an area; rather, different cultural and natural areas have different capacities. Instead, research has leaned more in the direction of normative values in understanding the needs of different types of users. Normative approaches provide information on specific user groups about appropriate use conditions and levels of impacts related to individual activities. In doing so they provide information (either qualitative or quantitative) which may be used by natural resource managers to establish management standards (Shelby and Vaske 1991). For example, it is not necessarily acceptable to suggest that 413 people be allowed to use a park over the course of a weekend. Although many parks and protected areas still maintain a numerical limit in controlling numbers, it becomes the task of the park manager to know the levels of expectations, satisfaction, dissatisfaction, and crowding of different types of users (i.e. motorboaters are likely to be able to withstand more use of the resource than canoeists, while canoeists would probably perceive the appearance of motorised craft as a threat to their experience; whereas the reverse might not be true).

The job of managing services and activities at a site, therefore, becomes a significant task. Park personnel must be receptive to queues – not only from the physical resource base (e.g. plant trampling and garbage) but also from visitors – when establishing regulations of where and what people can do. Pitt and Zube (1987) illustrate that once a resource manager has determined that the implementation of some form of recreational use limitation is necessary any of three overlapping courses of action needs to be considered.

## Site management techniques

Site management techniques focus on improving the environment's ecological capacity to accommodate use. This involves surface treatments (soil management) designed to harden the site where use occurs, and includes approaches that channel circulation and use into more resilient parts of the environment. Also, capital improvements may be developed in underutilised portions of the environment to draw people out of overused areas.

## Overt management approaches

Overt management approaches aim at direct regulation of user behaviour. They take several forms:

1   Spatial and/or temporal zoning of use (decreasing conflict of incompatible uses such as cross-country skiing versus snowmobiling).
2   Restrictions of use intensity (decreasing the number of users in the environment through the closing of trails).
3   Restrictions on activities/enforcement of user regulations.

## Information and education programmes

An alternative to heavy-handed overt methods:

1   Informing users about the recreational resource, and current levels of use.
2   Making the users more sensitive to the potential impacts their behaviours might have on the environment.
3   Giving the manager and the users a chance to exchange information concerning

user needs and management activities (e.g. brochures to describe entry points to users and usual intensity of use of different trails in order to distribute users more widely).

The regulation of visitor behaviour is a common approach to addressing management problems at recreation sites (Frost and McCool 1988). Such regulations often go beyond prohibitions on litter, alcohol, noise, and so forth, and directly restrict what tourists can do at a site, where they may go, and how many may be in an area at a certain time (overt management approach). Therefore, a tourist who wants to maintain a high degree of internal control might perceive the level of regimentation as too high within a certain opportunity, thus eliminating that alternative from consideration. (There is more in Chapter 6 on regulation as it applies specifically to tourism.)

In a study of Glacier National Park, rangers were given the task of managing visitors to this attraction, as well as protecting the eagles as an endangered species (each autumn recreationists come to view feeding bald eagles). Restrictions on use included prohibitions against entry where the eagles congregate, restrictions on automobile movement and parking, and close-up viewing available only at a bridge and a blind, but only with the accompaniment of a naturalist (acting as an interpreter and distributing brochures to visitors). With this in mind, the goals of the research were to understand how visitors responded to the current level of restrictions on behaviour, and how such factors as knowledge of the rationale for restrictions influenced these responses.

The authors discovered that 88 per cent of the visitors said they were aware of the park's restrictions, and that almost 90 per cent of these visitors felt that such restrictions were necessary, with only about 3 per cent feeling that they were not. Of the visitors who were aware of restrictions, 56 per cent felt that these had no significant influence on their experience, almost 32 per cent felt that restrictions facilitated their experience, and 12 per cent felt that restrictions detracted from the experience. When such restrictions were correlated with the concept of protecting the eagles, results indicated that visitors overwhelmingly support closures that minimise negative impact on eagles. Only 4 per cent of visitors perceived the opportunity to view eagles as a higher priority than eagle protection.

This study illustrates that visitors may have prior expectations for a certain degree of social control. The authors felt that visitors were likely to view management actions as acceptable and the regulations as enhancing attainment of certain outcomes, such as learning about nature (Frost and McCool 1988). Visitors who viewed the restrictions as unacceptable may ultimately be displaced. In addition, visitors were further impressed because they knew where and why closures and restrictions applied. This fact verifies the importance of the interpretive programme as a complement to management actions that regulate visitor behaviour. (More on interpretation in Chapter 6.)

Similar research has been conducted in New Jersey on the effects that ecotourists have on a variety of bird species in this region. Burger et al. (1995) report that birds are not consistent in their responses to human intrusions, and identify ecotourists as having the potential to disturb birds at all times of the year. This, according to the authors, is a result of the fact that ecotourists are interested in the breeding, wintering, and migration patterns of birds. For this reason, they have the potential to interrupt incubation, scare parents and chicks from nests, disturb foraging, disrupt the prey-base, force birds away from traditional habitats such as beaches, forests and open fields, trample vegetation and overuse trails. These authors felt that ecotourists and birds can coexist but only as the result of careful management of the resource, where each setting and species demands careful study and monitoring. They suggest the employment of the following measurements (Burger et al. 1995: 64):

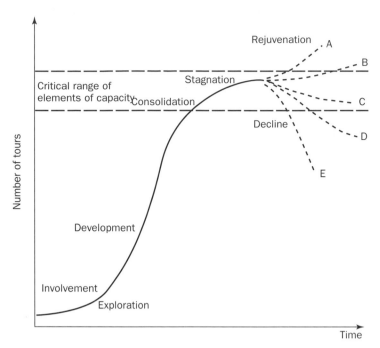

**Figure 4.2** The tourist area life cycle

*Source*: Reprinted from Butler (1980)

1  *response distance*: the distance between the bird and the intruder at which the bird makes some visible or measurable response;
2  *flushing distance*: the distance at which the bird actually leaves the site where it is nesting or feeding;
3  *approach distance*: the distance to which one can approach a bird, head-on, without disturbing it; and
4  *tolerance distance*: the distance to which one can approach a bird without disturbing it, but in reference to passing by the bird tangentially.

One of the most notable uses of carrying capacity in the tourism literature was developed by Butler (1980), who modified the product life cycle concept to apply to the life cycle of tourist destinations (Figure 4.2). Butler's basic premise was that increases in visitation to an area can be followed by a decrease in visitation as the carrying capacity of the destination is reached. Destination areas are said to undergo a fairly uniform transformation over time, from early exploration and involvement through to consolidation and stagnation, as the structure of the industry changes to accommodate more visitation and competing resorts. The implications of this research are such that planners and managers need to be concerned with any sustained decline in the ecological quality of the destination, as this will ultimately spell the demise of the development due to waning attractiveness to the tourist. This is a good example of a conceptualisation that applies to the social, ecological, and economic implications of tourism in a particular destinational setting.

More recently, researchers have focused on deriving empirical measurements of this evolution of a destination, especially island environments (Meyer-Arendt 1985; Cooper and Jackson 1989; Debbage 1990; Weaver 1990). The utility of the life-cycle concept has implications in delineating carrying capacity limits, and the social and environmental complications of 'overusage' in tourism destinations. Clearly defining the nature and characteristics of use of these areas must be a priority.

The Galápagos Islands of Ecuador is a case where carrying capacities have been considered as a means by which to control impact through the limitation of numbers of tourists on a yearly basis. The problem identified in the Galápagos is that despite the limitations on numbers visiting the islands, visitation annually increases beyond the limits set by management personnel because the economic impact of tourism is seen as the solution to the economic woes of this developing country, despite the efforts of researchers who recognise that the integrity of this precious global resource is in jeopardy owing to the inability of government to limit numbers. De Groot (1983) and Kenchington (1989) call attention to the fact that:

1   Patrol boats do not always control tourism numbers on the islands effectively.
2   The official limit of 90 tourists on an island at a time is often overlooked.
3   The number of tourists is still increasing. Total visitation has not been, but should be, kept under control.

These researchers suggest that tourism numbers have been controlled ineffectively and inappropriately through airport capacity limits rather than by limits set in accordance with ecosystem sensitivity defined by park planning and management. Thus, even in a well-known and highly significant area, problems of overuse and visitor management still arise. Wallace (1993) feels that it is the growth of the private sector which has been instrumental in dictating the course of action in the Galápagos. Park officials have found it difficult to enforce levels of acceptable use, zoning, and the distribution of permits owing to under-staffing and other, broader, political issues. The result is that park managers do not feel as though they are in charge of the operations of the park.

## CASE STUDY 4.1

### Ecotourism in the Galápagos Islands

Located 1,000 km off the west coast of South America, and straddling the equator, the Galápagos Islands which are comprised of some 120 islands, are perhaps one of the most intriguing ecotourism destinations. This stems from the rich history of the islands, which played host to the English naturalist Charles Darwin on his epic journey aboard HMS *Beagle* from 1831–36. Darwin and Alfred Russel Wallace (who was on the other side of the world in the Malay Archipelago) both almost simultaneously developed the theory of natural selection. The observations which Darwin made in the Galápagos Islands, particularly on the different species of finches, were instrumental in his formulation of the theory. Due to the international significance of Galápagos, based on Darwin's travel, the archipelago was designated as a wildlife sanctuary in 1935, but this was not enforced until much later (1959, a hundred years after the publication of *Origin of Species*). Twenty years on, the islands were classified as a UNESCO World Heritage Site in 1979; and a marine park was established in 1986. Throughout the late 1960s, 1970s, and 1980s, tourism steadily grew in Galápagos, consistent with the international growth of ecotourism, to the point where researchers began to call attention to the seriousness of environmental impacts from tourism on the region.

De Groot (1983) was one of the first to describe the negative impacts from tourism on the Galápagos in the academic literature. Among his many observations as a guide over a

*continued*

two-year period, he felt that the system of two guides which had developed in the Galápagos was ineffective. The educated, bilingual naturalist guides he felt were valuable, while the auxiliary guides who are not required to have extensive training and speak multiple languages often do not follow regulations. In the late 1980s, Kenchington (1989) noted that tourism to the islands was doubling every five years. This level of growth was encouraged by the Ecuadorian government, who refused to set limits on growth because of the economic infusion created by the tourism industry. This led to the development of two types of tourism in Galápagos: one which places heavy emphasis on nature and natural resources (ecotourism); and the other which is poorly organised, basic, little regulated, and run by individuals who have little respect for the biodiversity (mass tourism). This latter form of tourism in particular is placing a great deal of strain on the natural history of the islands.

The main reason for the escalating numbers of tourists is socio-economic. With a 40 per cent poverty rate in Ecuador, the Galápagos Islands have emerged as a gold mine for Ecuadorians. Ecotourism is thus seen as one of the strongest growth industries in the country. In 1997, the Galápagos provided income for an estimated 80 per cent of people living on the islands. Although dated, 1991 figures illustrate the value of tourism to the Galápagos to be about US$32 million, which had jumped to US$60 million by 1996. However, the open-entry philosophy of Galápagos tourism has already proved to be problematic. Price competition has led to a transfer of economic returns to foreign companies rather than to local people. Cruise ships, which are largely owned by foreign interests, are making the most money with very little multiplier effect. Furthermore, the jobs required to support ecotourism call for a high level of skill. Naturalist guides, as noted, must be able to speak two or three languages and have specialisation in natural history. This has proved difficult to attain for Ecuadorians and thus created the demand for guides from other countries. These issues, coupled with the development of two airports in the region, have paved the way for unprecedented numbers of tourists (approximately 80,000 tourist arrivals this past year) to help make these new developments cost effective, when many experts in the past suggested that half that number was an appropriate carrying capacity for the islands.

More specifically, a number of environmental problems from tourism have been identified by many authors. These include albatross, sea lions and turtles swallowing plastic bags (thinking they are jellyfish), the harvest of black coral for souvenirs, garbage found throughout many islands and coves, the introduction of sport fishing as an economic alternative under the guise of catch-and-release conservation, the introduction of non-native species, distress to nesting bird colonies, erosion and over-use of some trails. The human impact is also a concern, as suggested by Honey (1999), who observes that the annual growth rate of humans on the island is sometimes as high as 10 per cent, all of whom occupy 3 per cent of the archipelago. Coupled with a culture of diminishing respect for the natural history of the islands, the increased demand for new infrastructure is indeed significant. Most pressing, however, according to Honey (1999) and supported by Atwood (1984), is the introduction of non-native species which have both direct and indirect effects on plants and animals which have no natural defences (the island dilemma). In 1984, Atwood notes that controlling these animals (feral pigs, rats, goats, burros, dogs and cats) cost US$72,000 per species per year.

Some authors have noted that there is cause for optimism in this region. The development of a special law for the conservation of the Galápagos, with representatives from government, NGOs, industry and tourism, was designed for the long-term protection and profitability of the archipelago. Honey (1999) reports that the legislation is designed to support residents, stabilise populations on the islands, set aside an additional 2 per cent for human settlement, set aside more land for conservation, extend the zone of protection in the ocean, and ban industrial fishing for specific species. Although the law does not regulate tourism, it does limit tourism infrastructure (but not tourism numbers); it gives rights to local people to become involved in the tourism industry and stresses the importance of environmental education in schools. The park entrance fee of US$100 is one of the highest anywhere in the world, and this money, along with funds donated to the conservation fund, has resulted in substantial funds for many park programmes. In addition, the recent certification of five of the 20 large tour vessels that carry passengers among the Galápagos Islands, under the SmartVoyager programme, has provided more optimism for the region. This voluntary programme, which is a joint venture between the Rainforest Alliance and the Corporation for Conservation and Development (CCD), an Ecuadorian group, is designed to approve tour groups that tread lightly on this ecosystem. Boats are evaluated on a number of bases, including wastewater, fuel, docking, and minimising the introduction of foreign species.

After what might be viewed as an extensive period of disorganisation in tourism and other land uses, it appears as though the Galápagos Islands is now attempting to develop policy and to plan for the future. It remains to be seen how successful this development will be.

Websites on Galápagos ecotourism:

http://www.columbia.edu/cu/sipa/PUBS/SLANT/FALL98/p26.html
http://www.ecotourism.org/observer/062001/certification.asp
http://www.american.edu/TED/GALAPAG.HTM
http://www.rainforest-alliance.org/news/archives/releases/sv-tours.html

The general nature of Butler's model has ensured its applicability in just about any tourism situation. The rather unfortunate reality of his conceptualisation is that in many ways the life-cycle concept occurs today in much the same way as it did in the early 1980s when it was conceived (i.e. carrying capacities are still exceeded, large hotels still extract as much out of the destination as possible). Given the preceding discussion on sustainable tourism and AT, it is worth while to attempt to reconceptualise Butler's model while taking into consideration how such a cycle would, or rather should, proceed under ideal hypothetical sustainable tourism conditions. Figure 4.3 attempts to do this, emphasising the relative importance of economic, social, and ecological variables in establishing reasonable and long-term levels of carrying capacity within ecotourism destinations. The model illustrates that destination areas will respond to the competing economic and social and ecological demands in ways that respect the integrity of the resource base and local inhabitants. The overall level of visitation is intentionally kept below the identified level of acceptable use, over the long term, with potentially minor increases in use consistent with the ability of the environment to absorb such increases. Price mechanisms would therefore be implemented to ensure that acceptable financial gains are realised from the enterprise.

A logical manifestation of the carrying capacity concept has been the development of a number of preformed planning and management frameworks designed with the purpose of matching visitor preferences with specific settings in parks and protected areas. The ultimate aim of such frameworks is the protection of the resource base, but also to ensure that people are able to enjoy their recreational experiences in managed settings. Examples of these models include the recreation opportunity spectrum (ROS), the limits of acceptable change (LAC), the visitor impact management (VIM) process, and the visitor activity management process (see Payne and Graham 1993 for a good description of these frameworks).

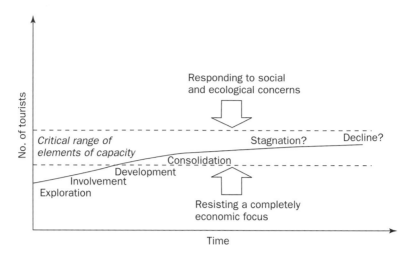

**Figure 4.3** A sustainable ecotourism cycle of evolution

Despite the relative success of these models in the realm of outdoor recreation management, there has been only a gradual use and acceptance of these frameworks by tourism researchers. This has generally been the result of the fact that these frameworks have not been developed specifically for tourism. In response to this fact, Butler and Waldbrook (1991) adapted the ROS into a Tourism Opportunity Spectrum designed to incorporate accessibility, tourism infrastructure, social interaction, and other factors into the planning and development of tourism. More recently, this framework has evolved into ECOS, or the Ecotourism Opportunity Spectrum (Boyd and Butler 1996). This model incorporates access, other resource-related activities, attractions offered, existing infrastructure, social interaction, levels of skill and knowledge, and acceptance of visitor impacts, as the means by which to plan and manage ecotourism *in situ*. Conversely, Harroun (1994) outlines the applicability of both the VIM (Loomis and Graefe 1992) and the LAC as a means by which to analyse the ecological impacts of tourism in developing countries. The main focus of these frameworks is to prompt decision-makers to ensure that an acceptable management framework is instituted prior to the tourism development process.

The environmentally based tourism (EBT) planning framework of Dowling (1993), however, is one such model that was developed specifically for tourism (instead of, for example, outdoor recreation). This model is grounded in the environmental disciplines and recognises that sustainable tourism planning can be accomplished only through a strong linkage between tourism development and environmental conservation. The EBT determines environmentally compatible tourism through the identification and linking of (1) significant features, including valued environmental attributes and tourism features; (2) critical areas, those in which environmental and tourism features are in competition and possible conflict; and (3) compatible activities, which include outdoor recreation activities considered to be environmentally and socially compatible. The EBT is based on five main stages and ten processes (Figure 4.4).

In general, the objectives stage of the model is important in that it involves the setting of the parameters of the study through discussions with government, local people, and tourists. It also involves consideration of existing policies affecting the study region, and the relationship between use and supply as they relate to tourism. In the second stage of the model both environmental attributes (abiotic, biotic and cultural features) and tourism resources (attractions, accessibility, and services) are assessed and integrated into a categorisation of

**Figure 4.4** The environmentally based tourism planning framework

*Source*: Dowling (1993)

sites. In the third stage, an evaluation of the significant features, critical areas and compatible activities, and the relationship of these to each other, is made and involves an overlay of both tourism and environmental attribute data. In stage 4 the identified significant features, critical areas, and compatible activities are matched with zones (i.e. sanctuary, nature conservation, outdoor recreation and tourism development), and nodes, hinterlands, and corridors identified at earlier stages of the project. The end-product of this stage is a map identifying the region's environmental units within the various zones. In the final stage the process is presented as part of an overall regional management plan. Discussions with resource managers are further required, and associated amendments to the plan in accordance with other land uses, in order for the tourism–environment plan to be implemented. The uniqueness of such a framework, as outlined by Dowling, is its environmental foundation, the incorporation of tourist and local opinions, the process of achieving tourism environment compatibility, and the fact that it presents itself as one of the only sustainable tourism planning models in existence.

## Assessment of ecological impacts

As identified in the previous discussion, a number of positive changes have occurred in tourism–environment research in the 1990s that relate to a better understanding of how the tourism industry has potential to compromise the integrity of the natural world. In particular, research has endeavoured to focus on the specific impacts of tourism in a number of different case studies, which is evident in both tourism and non-tourism journals (see, for example, Barnwell and Thomas 1995; Farr and Rogers 1994; León and González 1995; Price and Firaq 1996); and secondly to an interest in designing better ways of quantifying, and therefore assessing, such impacts. Goodwin (1995) writes that because tourists have been found to be bigger consumers of resources than local people at tourist destinations, careful environmental assessments of all new tourist developments are essential in documenting such overuse. Interesting, however, is that in an unrelated study on the perceptions of tourists, entrepreneurs, and locals in relation to who was responsible for the environmental impacts of tourism on Mykonos, Greece, Kavallinis and Pizam (1994) discovered that tourists felt that entrepreneurs and locals were more responsible for such impacts than themselves, and that local people considered themselves to be more responsible for ecological impacts, relative to the other two groups.

In an effort to better understand and control the effects of tourism development, environmental impact assessments (EIAs) have been employed to help quantify and qualify tourism impacts (Green and Hunter 1992). Although EIAs have been the topic of discussion for some 20 years in land use and planning, it is only more recently that they have been incorporated into the tourism development process. In general, the function of an EIA is to identify impacts that are of a non-monetary form, and thus enable developers to use resources efficiently in achieving a reasonably sustainable product over the long term. In addition, the importance of community well-being is an important consideration in the process of conducting EIAs in that there is a move to reconsider them not as a function of the built and natural environments but rather as a process that encompasses the needs of people in these settings.

Mitchell (1989) acknowledges that operationally EIAs sometimes fail, owing to limitations in researchers' basic knowledge (i.e. cause and effect relationships) of the system – the social and ecological conditions of the site – under study. He stresses the value of pure research in addressing this lack of information. However, before such 'hard' data are collected a substantial amount of 'soft' or qualitative data must be amassed in order to direct the more quantitative aspects of EIA (Green and Hunter 1992). These researchers emphasise the importance of methodological approaches such as the Delphi technique in allowing a subjective assessment of tourism developments by a series of stakeholders likely to be affected by the development. (The Delphi technique incorporates the use of successive rounds of a survey in gaining consensus on a particular issue. Surveys are repeatedly sent to experts in this process of reaching a consensus.) Typically, people involved in the Delphi include experts in various fields, such as planners, tourism officials, academics, engineers, environmental health officers, and so on, but also local residents and other affected stakeholders. Only after such information is collected, according to Green and Hunter, should more formal aspects of the EIA occur. It therefore gives many people the opportunity to aid in the process of identifying potential impacts that might not be recognisable to the planning team.

---

**CASE STUDY 4.2**

................................................................

### The fate of Mexico's Mayan heartland

Ceballos-Lascuráin (1996b) contrasts the development of tourism in the north and south of Quintana Roo, a state in Mexico's southeastern Yucatán. In the north, Cancún predominates as a mega-project of the 1970s, attracting more than 2,000,000 tourists per year. The social and ecological impacts of this development have been well cited, with beaches and lagoons being heavily polluted owing to a lack of appropriate sewage management and the creation of a marginalised economy between the few who are able to capture economic rewards from tourism and the many who have quite literally been displaced from traditional industries. Ceballos-Lascuráin writes that by contrast, the less populated region of southern Quintana Roo, without the beaches to support sun, sea, and sand tourism, has recently embarked upon a plan to develop a sustainable ecotourism industry (including archaeological tourism). Daltabuit and Pi-Sunyer (1990) however, report on the state of tourism development in neighbouring Chiapas (also part of the traditional Mayan heartland) and paint a very different picture of 'local' tourism initiatives. Using the example of the archaeological centre of Cobá, these authors illustrate that local people have had their land appropriated for tourism development without their approval, and been told that their opinions are meaningless in the tourism development process. These authors caution that the international community will have to recognise that tourism (in relation to proposals to further develop the Mundo Maya region for tourism purposes) is not a benign and positive force for conservation and cultural heritage. Rather, it occurs as a function of political and economic factors, in a top-down fashion, that is often at odds with the needs of local people.

---

There can be little question that significant gains can be made by incorporating local people into the planning and development of an ecotourism project. In the delivery of other recreation products, the term 'positive affect' refers to the need that people have to exert some influence or control, however minor, in shaping their recreation experiences (e.g. through suggestions, comments, and so on). Good recreation programmers therefore know that positive leisure experiences are in many cases contingent upon the satisfaction that people get in being asked to comment on various ways of offering a programme in which they have been or will be participating (see Chapter 7). In tourism development, positive affect provides some theoretical basis for allowing local people to control, at least in part, the events that unfold within their community. As we know, much tourism development in the past has occurred without the consent of the vast majority of community members (see Chapter 8 for more discussion on community development in tourism).

## Conclusion

................................................................

Tourism research has been successful in identifying a range of social and ecological problems brought about by the tourism industry. Future research must endeavour to explore new avenues and approaches to controlling tourism impacts. For example, past impact studies have usually concentrated solely on the effects that tourists have *within the destination*. Future macro-based studies may attempt to correlate the effects of an individual within the tourist region but also in light of the effects that person would also have had on their home environment. In this more global sense, the impact that a tourist has on the destination (the ecological footprint) could in fact be less severe than his or her impact at home. In addition, researchers may wish to develop impact budgets (a spin-off from the time budget methodology) which might be designed to document the real and perceived impacts that tourists themselves have on the resource base. The results of such a device may be used as an educational tool for tourists, hoteliers, and so on, in demonstrating the effects of tourism.

## Summary questions

1   Discuss ways in which the tourism industry can have a transforming and dislocating impact on the social fabric of host communities.
2   What is carrying capacity, and how are social and ecological carrying capacities separate but related?
3   List some examples of different site, overt, and educational management styles.
4   What are the implications of Butler's tourist area life cycle for tourism development? Why is carrying capacity such an important aspect of this model?
5   What is a preformed planning and management framework, and why are these reported to be better at managing human impacts in natural areas than traditional carrying capacity techniques?

## 5

**The economics, marketing, and management of ecotourism**

In this broadly based chapter the economics, marketing and management of ecotourism are discussed using a variety of examples and research studies in the literature. The section on economics examines the current predisposition of researchers to predict the global economic impact of ecotourism, in addition to a discussion of leakages and the multiplier effect, revenues in parks, and the economic value of land. The section on marketing evaluates the potential of marketing in the ecotourism field, but also addresses the need for such studies to be accurate in their projections of the market, now and in the future. The final section on management discusses a number of issues thought to be relevant to the field, including privatisation, not-for-profit organisations and NGOs.

## The economics of ecotourism

The World Tourism Organization projects that international tourism arrivals will have grown from the 593 million figure of 1996 to over 1 billion by 2010. At the same time earnings from international tourism are expected to climb from the 1996 figure of US$423 billion to US$1.5 trillion in 2010 (Luhrman 1997). Such is the magnitude of the tourism industry, which by many accounts is reported to be the world's largest. With continued growth at the rate identified above, the general trend does not appear to be faltering.

In a review of the literature on the economic impact of the global ecotourism industry, there is the feeling among some researchers that ecotourism is expanding even faster than the tourism industry as a whole (see Lindberg 1991; McIntosh 1992; Hawkins in Giannecchini 1993), with upwards of 20 per cent of the world travel market as ecotourism (Frangialli 1997). Others, however, are more sceptical (Horneman and Beeston in Tisdell 1995) and feel that it is more logical to view ecotourism's growth in a site-specific manner than as a general overview. Jenner and Smith (cited in Goodwin 1996) estimate that ecotourism had a global value of $4 billion in 1980, $5 billion in 1985, and $10 billion in 1989. They forecast ecotourism's value to be $25 billion in 1995, and $50 billion by the year 2000; whereas Tibbetts (1995–6) illustrates that of the $2 trillion that tourism generates annually, $17.5 billion is from ecotourism. Perhaps more realistic is the WTO's recent claim that ecotourism constitutes 2–4 per cent of global tourism [WTO (2002) International Year of Ecotourism launched in New York (online): http://www.world-tourism.org/newsroom/Releases/more_releases/january2002/launch]. Such statistics indicate the disparate views that exist on the economic impact of ecotourism. One of the main reasons for such a discrepancy in quantifying ecotourism is the lack of a clear-cut definition of the term (Hawkins in Giannecchini 1993).

## The flow of local money

Foremost in the minds of many local, regional, and national bodies charged with the responsibility of tourism development is the importance of earning money. However, all such regions are not created equal in terms of their ability to generate and keep money within the economy. Simply stated, although tourism will always create an infusion of money into local economies, the amount of money that stays in that economy is subject to a number of factors. In evaluating the impact of money on the economy, economists must understand the multiplier effect and the associated concept of leakage.

In general, as new money from tourism enters a local economy it changes hands many times, resulting in a cumulative economic impact that is greater than the initial amount of tourist expenditure. More specifically, direct income or first-round income is the amount of that spent money that is left over after taxes, profits and wages are paid outside the area and after imports are purchased (Getz 1990). The money that remains after these leakages is referred to as secondary income, which circulates successively through the economy creating indirect income and induced income, again with various amounts of leakage occurring. National multipliers tend to be highest and, according to Bull (1991), have ranged from 2.5 in Canada to 0.8 in Bermuda and the Bahamas. Most of the less developed economies, especially those island economies, tend to be lower because of the high level of leakage. It follows that even in the developed world, the multipliers of small towns and counties are typically low (in the neighbourhood of 0.25). Leakage rates can be determined by assessing the percentage of money that flows out of the economy (see Figure 5.1).

Clearly, those regions or localities that can minimise the amount of money leaving their economy will have more of the initial expenditures left to circulate through the economy. Those regions or localities that rely upon resources from outside the region or country – human, physical, and monetary – will no doubt suffer by having to pay for their services.

**Figure 5.1** Imports leading to leakages

*Source*: Goodwin (1995)

The reverse also holds true in that those regions that can sustain a tourism industry based on the resources that they have at hand will potentially be able to prosper under these conditions. This form of leakage is referred to as import substitution. It is an important concept in the context of ecotourism and sustainable tourism because there is much evidence pointing to the fact that tourism, for example in the less developed countries (LDCs), has been hampered by the fact that management control of the industry lies in the hands of external, multinational interests. Hotels, car rental agencies, restaurants, and airlines, all the big money-makers in the tourism industry, are quite often owned by companies that reside outside the destination region, and the destination region relies upon these to export its product through tourism.

## Revenue and parks

Tourism is inherently a private-sector activity that capitalises on a market for the purpose of making a profit. A conflict emerges when a profit-motivated enterprise relies on the provision of supply that does not necessarily advocate the same market philosophies. Parks and protected areas, as public entities, provide the cornerstone for the ecotourism industry. In many cases there is debate over whether parks should be operated more as a business in response to shrinking public budgets. Conventionally the management of parks has not been subject to the same market principles and philosophies as the private sector. While it is generally accepted that ecotourism in protected areas has positive economic spin-offs (e.g. direct employment, both on- and off-site, the diversification of the local economy, the earning of foreign exchange, and the improvement to transportation and communication systems), there are also associated negatives, including the lack of sufficient demand for ecotourism, which could result in the draining of badly needed funds; the fact that ecotourism may not generate local employment opportunities; the fact that leakages may be quite high, as they are in many small and developing regions; and the fact that it may not be socially and economically acceptable to charge fees in parks.

The preceding discussion alludes to the fact that one of the most significant issues facing parks and protected areas today is the means of attaining funds for their operation – indeed, in many countries, for their survival. Many parks agencies that have historically relied on certain specific means by which to support their parks and park services have had to consider diversifying in order to maintain good-quality programmes and infrastructure. Sherman and Dixon (1991) illustrate five main ways in which to gain revenue from nature tourism, including:

- *User fees*. These are usually a reflection of the public's willingness to pay, and in recent years have altered into more of a two-tiered or multi-tiered system with a differential scale of fees, the fee varying according to whether the visitor is a resident or a foreigner.
- *Concession fees*. In the case of government, fees are charged to private firms who provide tourists with goods and services (guiding, food, etc.).
- *Royalties*. Souvenir and T-shirt sales provide a good basis of this type of revenue, which is given to the agency as a percentage of the revenue made on the items.
- *Taxation*. Sales tax, hotel tax, and airport tax are examples.
- *Donations*. Tourists can be prompted to contribute money in an attempt to address a local problem (lack of resources or money for endangered species) and, in the process, aid in the management of a protected area.

In the case of the Monteverde Cloud Forest Reserve of Costa Rica, the issue of park entrance fees has been widely discussed in the context of the viability of this reserve. According to Aylward *et al.* (1996), in the early 1970s visitors, regardless of their origins, were asked

**Table 5.1** *Guiding principles for fee policy in nature-based tourism*

| Principle | Rationale |
| --- | --- |
| Fees supplement but do not replace general sources of revenue | Even for heavily visited sites, fee revenue rarely covers total costs, especially capital costs. Heavy dependence on fee revenue reduces visitor diversity and the scope of attractions that can be offered. Yearly fluctuations in fee revenue make fees an unstable income source |
| At least a portion of fee revenues should be set aside ('earmarked') for sites that generate them | Earmarking increases management's incentives to set and collect fees efficiently. Visitors may be more willing to pay fees if they know that fees are used on-site |
| Fees should be set on a site-specific basis | National guidelines specify fee objectives and policies, yet management goals and visitor patterns vary across nature-based tourism sites, requiring local flexibility in assessing the type and amount of fee |
| Fee collection is not justified at all sites | Fees are not cost-effective at places with low visitation demand and high collection costs |
| Fee systems work best when supported by reliable accounting and management | Administrative decisions about fees require acceptable data on costs and revenues of providing NBT for different sites and activities |

*Source*: Laarman and Gregersen (1996)

to pay a fee of approximately US$2.30 to gain entrance to the reserve. However, owing to the increasing levels of visitation, and the demands placed on the reserve, a new fee structure had to be developed. In 1995 fees were restructured as follows: <$1 for Costa Rican students; $1.50 for Costa Rican residents; $4 for foreign students; $8 for non-package tour foreigners; and $16 for foreigners on tours (ibid.). Such fees have been instrumental in enabling this private reserve to become more economically self-sufficient.

Tisdell (1995), however, suggests that there are some inherent limitations to the implementation of fees in financing ecotourism developments, including the possibility that few people may visit the site and/or if the park is located in peripheral areas. As alternatives, managers might elect to make visitors purchase permits from park offices, require tourist operators to pay visitor fees (as is the case in many destinations), or erect automatic ticket machines in car parks and trail heads – an approach that has worked well in Pacific Rim National Park, Canada. Laarman and Gregersen (1996) and Steele (1995) concur that pricing holds tremendous power in providing greater efficiency and sustainability in ecotourism, but is seriously neglected in public policy. These authors identify pricing objectives, pricing strategies, and categories of fees in arguing that the user-pays principle and the removal of free access to public lands are perfectly logical today as a means by which to recover costs and, indeed, make money. While it is not necessarily the goal of publicly run natural areas to 'profit', it is for community-run organisations and of course private enterprises. Laarman and Gregersen offer some guiding principles for fee policies in nature-based tourism, as shown in Table 5.1.

Laarman and Gregersen argue that pricing objectives can be many-sided, and administrators are constantly challenged to set fees in accordance with the resource conditions of the park, the needs of the park staff, and the needs of visitors. In Table 5.1 the guiding principles are varied and range from a relationship of fees to general sources of revenue, fees for certain sites, fees only for certain sites, and the management and accounting of fee systems. In some park systems fees that are generated in each of the individual parks go back into a general operational account. The positive spin-off of this is that the money that is generated for this account aids in the maintenance of parks in the entire system. The negative element is that those parks doing the best job (either because they have better administrators or because the park simply generates more visitation) do not get the opportunity to utilise directly for their own purposes, the money that they generate. This type of fee philosophy may further decrease the motivation of those working in the money-generating parks, so that they become less conscientious, and may lead those in the money-losing parks chronically to rely on the money generated in other areas.

Steele (1995) illustrates that a policy allowing for open accessibility to ecotourism sites leads to certain economic and ecological inefficiencies. Economic inefficiencies occur in the sense that if sites allow free entry they lose the rental value of the resource. Agencies must also be wary of excess demand where sites are left vacant during certain seasons of the year, and also of costs related to congestion where tourists impact on each other owing to overcrowding (congestion costs are found to reduce the profit per tourist by lowering tourist demand and raising marginal costs for the supplier). Ecological inefficiencies include consideration of carrying capacities and an analysis of the total volume of tourists and the damage done per tourist. By distributing tourists appropriately in space and time, some natural areas, according to Steele, have been able to increase the numbers to the natural area and reduce the overall impact of these tourists. Also of importance to Steele is the choices that land managers have with respect to pricing controls and quantity controls (limits on numbers of users, which are employed more often than pricing controls). This author suggests that the use of variable tariffs or tiered pricing is an efficient way to increase revenue (see the example of Monteverde, pp. 85–6), where foreigners are charged a higher entrance fee than locals. The most oft-quoted example of this is the Galápagos Islands, where foreign tourists are charged the equivalent of US$100 and local tourists for less.

McFarlane and Boxall (1996) write that, historically, many public wildlife agencies get their funds from hunting and fishing licences and general tax revenues. However, owing to the recent decline in many consumptive forms of outdoor recreation (e.g. hunting), and budgets, financial resources for conservation initiatives are dwindling. Their research of 787 birdwatchers indicated that this group shows great promise in supporting conservation efforts in a number of ways. While committed and experienced birders could help to identify wildlife management issues and participate in fieldwork with various agencies, the less specialised birders could aid by improving bird habitat in their backyards and through the contribution of funds. The research demonstrates that conservation agencies must diversify and include innovative schemes to capture the interest and support of the birding population.

## The value of land

In determining the appropriate use of a resource area, government and other decision-makers must determine the most financially and socially acceptable use of land. Decisions must be made whether or not (and how) to develop dams, parks and protected areas, mining, or forestry, all of which have certain associated costs and benefits which will most certainly differ in different settings. Land planners and developers have begun to recognise that destinations are increasingly demanding business ventures that will add value to raw materials (McIntosh 1992; Theophile 1995), instead of taking it away. While many of the large tourism developments of the 1950s, 1960s, and 1970s occurred with only the financial motivations in mind, these days, in most cases, such large endeavours are undertaken with both ecological and socio-cultural impacts in mind (see Chapter 4 for information on impact assessments). Given ecotourism's attractiveness as a sustainable tourism option, it is critical to have development ideas accompanied by appropriate social and ecological measures.

**Figure 5.2** Categories of economic values attributed to environmental assets

*Source*: Munasinghe (1994)

One of the most effective ways in which to decide on how best to develop or not develop an area of land is through the implementation of a cost–benefit analysis, which includes easily quantifiable costs and benefits (such as start-up and operating costs) and those which are not easily quantified, such as impact assessments (see Chapter 4). Despite the difficulty in doing so, researchers have endeavoured to place values on ecological and social settings. In the case of the environment, Munasinghe (1994) has categorised economic values attributed to ecological resources by examining the use and non-use values of assets. He suggests that the total economic value (TEV) of a resource is based on its use value (UV) and non-use value (NUV), which are further broken down in Figure 5.2. In the figure, Munasinghe shaded the option values (an individual's willingness to pay for the option of preserving the asset for future use), the bequest values (the value that people derive from knowing that other people will benefit from the resource in the future), and the existence values (the perceived value of the asset) as a caution to the fact that they are all quite difficult to define. From here, analysts may use a number of non-market valuation techniques to quantify the above values (Munasinghe 1994; see also De Lacy and Lockwood 1992), including the travel cost method and the contingent valuation method.

The travel cost method, for example, has been employed by Menkhaus and Lober (1996) in Costa Rica as a means by which to estimate the values of ecotourism to tropical rainforests, based on a sample of US travellers. This valuation method, according to the authors, 'estimates ecotourism benefits of a protected area based upon observed travel expenses by visitors to an area' (ibid.: 2). These estimates are then used by decision-makers to address issues related to entrance fees, competing land uses, and so on. The argument for the use of such a method is that certain goods such as parks are not effectively valued on the basis of entrance fees alone; individuals, it is felt, would be willing to pay much more to experience the good. The measure of the value of a good such as ecotourism to a protected area is represented by the consumer surplus. In their study Menkhaus and Lober estimated the consumer surplus to be approximately $1,150, which represents the average annual per person valuation of the ecotourism value of protected areas in Costa Rica for the sample. These authors estimated the annual ecotourism value to Costa Rica's rainforests by this population at $68 million, which is determined by multiplying overall US ecotourist visitation to Costa Rica by the per person consumer surplus.

An example of the contingent valuation method is found in the work of Echeverría *et al.* (1995) in their evaluation of the Monteverde Cloud Forest Reserve of Costa Rica. This technique, otherwise known as the direct or survey method, is an approach 'in which individuals are asked directly about their willingness to pay for some improvement in a resource or for additional use of a resource' (Smith 1990b: 249). This can take the form of an assessment of the willingness of tourists to pay for habitat protection. Echeverría *et al.* sought to address these non-market values of the visitors to Monteverde in demonstrating that, on average, foreigners were willing to pay $118 of their own income to secure reserves such as Monteverde. This figure is comparable to the willingness of British Columbia residents to pay money to increase the scope of the protected areas in that province (as identified in Chapter 3).

In valuing places like parks and protected areas (and tourism within such areas), it must be realised that there is a significant degree of grey area between the ideologies of business and those of conservation. Profit and preservation, therefore, are often construed as a partnership, according to Carey (in Giannecchini 1993), that is analogous to an arranged marriage, not one based on love. For example, Giannecchini notes that:

> Their [the tourism industry's] customary goal of quick optimum profits is in direct conflict with long-range goals of protection and conservation. This does not mean that the only, or even primary, relationship between the tourism industry and conservationists must be adversarial. But it does mean that whatever laudable, environmentally sound policies and goals the industry articulates, they will remain subsidiary to the demand for profit. Therefore, if the tourism industry becomes the principal force in the development of ecotourism, it will almost certainly be detrimental to long-range environmental concerns.
>
> (Giannecchini 1993: 430)

A key means by which to prove the value of ecotourism (in economic terms) to those advocating other land uses has been economics. Researchers have looked at this type of tourism as a way to demonstrate that in light of other competing types of land use, ecotourism presents itself as an effective form of land use. For example, McNeely (1988) estimated the economic worth of specific animal species in Amboseli National Park in Africa. He figured that each lion is worth about $27,000 per year, and each elephant herd $610,000 per year, which far exceeded the amount of money that could be gained through hunting these animals. Western and Thresher (1973) wrote that in Amboseli tourism could be much more environmentally desirable than agriculture, and in addition earn vastly more per unit hectare of land ($40 per hectare from tourism and $0.80 per hectare from agriculture).

This form of analysis brings to mind the ways in which people value objects such as wildlife. Clearly there is the economic focus, yet examples of this to some degree are required as a means by which to rationalise the existence or occupance of something, like ecotourism, as a new land use. Kellert (1987) suggests that attitudes towards wildlife as a commodity or utilitarian object became less widespread in American society between 1900 and 1975. He believes, however, that this attitude change has occurred in only some parts of society. Those who are elderly, rural, in a lower socio-economic category, and who live and work in resource-dependent occupations still harbour negative perceptions of wildlife. It is often difficult for those who are not in one of the aforementioned categories to conceive of such perceptions. Although the point is not made in Kellert's research, one wonders about the impact that urbanisation, education, and wealth have in shaping the more positive attitudes towards wildlife. Ecotourism is, or has been, conventionally considered an activity pursued by those who are better educated and wealthier than the 'average tourist' (as identified in Chapter 2), which would, according to the work of Kellert, lead one to believe that these individuals have a better chance of appreciating wildlife. Those who are providers of the ecotourism experience in the least developed countries – assuming that they are not part of the privileged elite of that country – must acquire a 'personal conviction that land managed for wildlife is land ultimately more satisfying, attractive and enjoyable for people' (Kellert 1987: 228).

## Marketing

Mahoney (1988) has suggested that tourism marketing fundamentally differs from the marketing of other types of products in three important ways: (1) tourism is primarily a service industry, where services are intangible, and quality control and evaluation of experiences are more difficult to envision; (2) instead of moving the product to the customer, the customer must travel to the product or resource; and (3) people usually participate in and visit more than one activity and facility while travelling. Therefore, tourism-related businesses and organisations need to cooperate to package and promote the tourism opportunities available in their areas.

The marketing of tourism products is strongly based on a firm understanding of the fact that the overall travel market is partitioned into selected market segments. Travel firms do not have the resources to tap the overall travel market, nor the inclination, owing to the magnitude of domestic and international travel. Instead, businesses target certain segments based on the product that they are selling and the needs and expectations of the group to which they wish to sell. Target marketing can occur in four main ways: (1) geographically, on the basis of geographical space; (2) demographically, based on age, gender, religion, race, etc.; (3) psychographically, based on individuals' lifestyles, attitudes, values, and personalities; and (4) benefits, which include an analysis of the benefits sought by tourists and the costs they avoid.

Marketers typically follow a systematic planning approach that enables them to focus on their organisational goals, and the specific needs of their clientele in association with various tourism products. A typical plan, based on the work of Lovelock and Weinberg (1984), is outlined in Figure 5.3. The first stage of the marketing plan involves the identification of the direction of the agency or organisation and the associated priorities that must be followed. This is followed by the definition of markets that will allow the organisation to achieve its chosen goals. Upon the completion of this stage, the organisation endeavours to examine fully the behaviour, needs, and characteristics of the market. Once the markets have been segmented, the organisation can develop specific strategies which must be tailored to each segment. There follows the implementation of a monitoring and evaluation component, with subsequent modification in the future.

Demand is very much an integral part of the relationship between the tourism product and the market, and hence Krippendorf (1987) feels that in the future tourism marketers must be more sensitive to the changing composition of the tourist population. He suggests that owing to the changing lifestyles, economic conditions, and demographic structure of travellers, 'the market is shifting from manipulated, uncritical "old tourists" to mature, critical and emancipated "new tourists"' (ibid.: 175). Sixteen years on, this trend has certainly taken shape.

As is the case with the conventional travel markets, marketing research in the area of ecotourism has begun to flourish. The work of Ingram and Durst (1987) stands out as one of the earliest studies in this area, through their development of a bibliography on marketing

| | |
|---|---|
| Define institutional objectives and set priorities | Objectives and priorities |
| Identify which market(s) within overall population is/are key to successful achievement of objectives | Market definition |
| Examine key market(s) to determine needs, personal characteristics, and current behaviour patterns | Analysis of target markets |
| On the basis of above analysis, further divide key market(s) into target segments that are sufficiently different to merit tailored strategies | Development of target market segments |
| Develop efficient marketing strategies targeted at each segment | Tailoring marketing strategies |
| Monitor results for each target segment and evaluate relative success of marketing strategies in efficiently achieving desired objectives | Monitoring and evaluation |
| On the basis of evaluation, make any needed modifications to (1) segmentation approach employed and (2) marketing strategies directed at each target segment | Modification |

**Figure 5.3**  The process of market targeting to achieve institutional objectives

*Source*: Lovelock and Weinberg (1984)

nature-oriented travel in developing countries and a subsequent publication that looked specifically at promotion of nature tourism in developing countries (Durst and Ingram 1988). From their research they concluded that there was a need to improve the operations of tourism offices in the developing world. They also found that although marketing is common in the developing world, countries that are not promoting their natural attractions are likely to miss the opportunity to capitalise on the growing ecotourism market.

Ryel and Grasse (1991) suggest that the two cornerstones to effective marketing in ecotourism are the attraction for tourists (which includes biodiversity, unique geography, and cultural history), and tourism infrastructure to support the industry. Given these criteria, agencies are free to undertake the necessary marketing requirements to attract tourists. These authors imply that it is important to attract or market to the 'right' clientele. This includes 'born' ecotourists (those who have a built-in predisposition towards nature and nature travel), and 'made' ecotourists, who can be identified as representing latent demand or those who are unfamiliar with this form of tourism but who can be attracted through effective marketing. Ryel and Grasse suggest that a basic approach to marketing the ecotourism product is through (1) the identification of the characteristics of a desired group, (2) appropriate advertising, (3) careful crafting of the advertising message, and (4) the development of a mailing-list. These authors further suggest that there is a definitive need to classify ecotourists on the basis of demographic, psychographic and geographical variables, for, as Mayo (1975) has suggested, demographics alone are not an effective means by which to segment the market. This fact has been further identified by Silverberg et al. (1996), who undertook an investigation into the psychographics of nature-based travellers. After identifying six factor dimensions on the basis of their survey of 334 travellers to the south-eastern United States (education/history, camping/tenting, socialising, relaxation, viewing nature, information), they concluded that psychographic research is an effective way for marketers to understand the nature-based traveller better.

In Australia, the Office of National Tourism undertook a market study of ecotourists in an attempt to understand the ecotourist market better and target ecotourism products more effectively (Commonwealth of Australia 1997). This research involved a number of focus groups involving ecotourists (actual and potential) from across the country. In general, the study found that ecotourists are seeking the following: (1) areas/attractions of natural beauty; (2) small groups away from crowds; (3) some level of interaction with the environment; (4) interaction with like-minded people (see also Fennell and Eagles 1990); (5) some degree of information and learning; and (6) fun and enjoyment. More specifically, the research uncovered three broad ecotourism market segments. These are as follows (see also Chapter 2 and the discussion on ecotourists):

1   *Impulse*. Characterised by nature-based day trips away from the main tourist destinations and mainly booked locally by both domestic and international tourists. The level of activity on these tours varies widely.
2   *Active*. Characterised by younger and middle-aged professionals who generally book in advance. There is a skew to domestic tourists, although there could well be potential for growth through international marketing, infrastructure, and product development.
3   *Personalised*. Essentially older professionals (or retired) who expect to be well looked after by the operator. This segment is skewed to international tourists who book overnight ecotours before arriving in Australia (Commonwealth of Australia 1997: 4–5).

In 1994, the Canadian provinces of Alberta and British Columbia published a comprehensive marketing study (ARA Consultants 1994) which focused on assessing current and future demand for ecotourism in these regions. The authors of the report sought information from a variety of sources in formulating their study, including a literature review of ecotourism, a travel trade survey, a general consumer survey of the residents of seven major metropolitan cities in Canada and the United States, and a survey of experienced North American ecotourists. Ecotourism was defined as 'vacations where the traveller would experience nature, adventure, or cultural experiences in the countryside' (p. 1–5), and hence treated cultural tourism, adventure tourism, and ecotourism as being synonymous. One of the main conclusions from this research was that 'the ecotourism market is definitely expanding to encompass not only those with specialist skills seeking strenuous adventure, but also clients seeking activities supported by amenities, a higher level of services, and requiring lower levels of specialist skills' (p. 6–1). This finding is no doubt a result of the methodology employed in the study. Other conclusions of the study are as follows: For the general consumer group, 77 per cent said they had an ecotourism vacation, with the remaining 23 per cent expressing interest in ecotourism travel. This led the researchers to conclude that from the general consumers of the seven cities surveyed, the actual market ranges from 1.6 to 3.2 million visitors, and the potential market is approximately 13.2 million ecotravellers. To two of Canada's most beautiful and 'natural' provinces, it is not inconceivable that travellers would involve themselves in any type of natural, adventure, or cultural experience. This probably extends to hard-path ecotourism trips right along the spectrum to bus tours of the city of Victoria.

The point to be made from this example is that researchers and practitioners must proceed cautiously in making inferences on the size and scope of the ecotourism market. Past research has concluded that ecotourism is a very small part of the overall tourism continuum. Although it is certainly growing, so are the ways in which to classify and define ecotourism, many of which are based on convenience and less on a thorough review of ecotourism and ecotourism types. The research makes the assumption that, apart from the involvement of cultural and adventure tourism types, experienced and inexperienced ecotourists are one and the same. In substantiating this point, Wight (1996) reports that, with respect to the experienced North American ecotourist sample,

> Because client lists were volunteered, it is not known to what degree those firms volunteering names are representative of the ecotourism travel trade. Consequently, it is also not known how representative the respondents are of all ecotourists. However, the sample does represent a large number of experienced North American ecotourism travelers.
>
> (Wight 1996: 3)

In this case the softest of the soft-path ecotourists (or culture and adventure tourists) are grouped with the hardest of the hard-path ecotourists, groups which are perhaps worlds apart in their attitudes, motivations, and benefits sought. This ambiguity is indicative of the activity packages offered by the tour operators from which the 'experienced ecotourism' sample was drawn. Ranked in order of number of packages offered, hiking was first, followed by rafting, canoeing, cycling, kayaking, horseback riding, and then wildlife viewing – activities which are more skill-oriented and less learning-oriented in nature (see Chapter 2 for a discussion on comparing adventure tourism). Consequently, it was found that water-based activities were important, but particularly important to the experienced ecotourist (not surprising, and due most likely to the types of ecotourism packages offered); that virtually all ages of adults are interested in ecotourism; and that interest in ecotourism is spreading to less educated groups.

The results of this study have an element of inclusiveness, indicative of the methodology employed. Accordingly, the usage of the term 'ecotourism' has spun out of proportion and been misused as an industry label to capture a larger percentage of the travel market (see Goodwin 1995), with the main implication of misrepresenting or 'watering down' the ecotourism market. For example, Thompson (1995) writes that a cruise aboard luxury liners, scuba diving, and helicopter sightseeing trips over Hollywood are all being touted as ecotourism. Clearly, in such cases ecotourism has been modified, repackaged, and 'mass' produced to the point where the line between what ecotourism is and is not, is quite blurred. If this is the approach by the industry, and if ecotourism is to become successful by conventional industry standards, it will have undermined exactly what it set out to accomplish in the first place. The misrepresentation of ecotourism was the source of a recent debate over the TRINET (the tourism research international computer link organised out of Hawaii). For example, Iverson (1997) writes that:

> When I sought to 'dive with the seals' in Kaikoura NZ I was attracted to a nice glossy brochure with eco-labelling. My accommodation host saw my interest and recommended another company, a fellow who 'started it here'. When I met the fellow with this beat-up 4 wheeler hauling his van of dive gear I was wondering if I made the right choice. I had a great time and discovered later that the company with the glossy brochure had been in trouble with the Dept of Conservation for exceeding their permits and was told that they would take 24 people out with 12 permits and put half of them in the water at one time.

Ryan (1997) follows up on this discussion by making similar observations on the state of affairs in New Zealand. He suggests that there are basically two types of ecotour operators: one that constructs a glossy pseudo front, and the other that is small in scale and that happily accepts size restrictions. Continuing the discussion, Oppermann (1997) felt that too many tourists are subsumed under the ecotourism label – a stance adopted in this book – and that researchers need to be wary of studies that use wide-ranging definitions of the term. In order to avoid the misrepresentation of the ecotourism 'industry', Mason (1997) suggests that a system based on self-regulation is suspect and fraught with abuse. He would rather see a form of policing and/or monitoring to avoid the unacceptable behaviour of those operating under free-market conditions. The problem to many marketers, operators, and agents has become so bad that they are refusing to use the ecotourism label because it has gained a poor image and means very different things to different people (Preece *et al.* 1995).

A further problem for regions in the development and marketing of an ecotourism product is the fact that there seems to be a feeling of unlimited assurance that ecotourism will ultimately resolve any tourism dysfunctions within a region. Tourism decision-makers and planners, perhaps on the basis of overly optimistic forecasts of management consultants, often inherently believe that their jurisdictions will be competitive in a highly competitive global ecotourism market. This has currently become an issue in Canada, where provinces that cannot hope to compete at the scale of other provinces – owing to their geological and natural features – have unbridled optimism for the sector. There is no question that eco-tourism can and will exist in these regions, but regions must be realistic about the success of ecotourism initiatives. The size and extent of the sector will be a function of the demand, but it does not restrict them from creating innovative schemes to attract ecotourists given the attraction base from which they have to work.

As a case in point, the province of Saskatchewan, in association with Industry Canada, has recently commissioned a study on the market potential of ecotourism in the province (Anderson/Fast 1996a, 1996b). The research was designed to gather valuable information from the travel trade (companies currently operating adventure, culture, nature, and ecotour trips) on the potential for Saskatchewan as an ecotourism destination. One of the survey questions sought industry response on nine potential ecotour packages to be offered in the province. In only one case did over 50 per cent of the tour companies sampled feel that they could sell the package to their clients. One of the most significant ecotour attractions in the province is the migratory and resident bird populations. However, in the case of the 'Birding in Saskatchewan' tour, only 43 per cent of the operators said that this tour appealed to them, and only 48 per cent said they could sell the package to their clients. The reasons cited for the lack of interest included 'wrong fit', 'too expensive', and 'not stunning enough'. Other findings of the study indicated that a lack of information on the province and lack of demand for Saskatchewan ecotourism products were two most frequently cited barriers confronting the ecotourism industry in Saskatchewan.

A report on the ecotourism opportunities for Hawaii's tourism industry (Center for Tourism Policy Studies 1994) put ecotourism and Hawaii into perspective by acknowledging the fact that Hawaii faces some stiff competition from already well-established global ecotourism destinations (i.e. Latin America, Africa, and Asia). They found that the main considerations constraining ecotourism in Hawaii could be summarised as follows (pp. ii, iii):

1   *Hawaii's physical characteristics.* Hawaii's isolation, limited size, and fragile natural environments create conditions which can lead to irreversible environmental harm from the overuse of its natural resources.
2   *Hawaii's culture and local lifestyles.* Native Hawaiian issues and the preservation of community values have become increasingly sensitive to land use decision-making.
3   *Hawaii's competition abroad.* Hawaii faces competition from a number of well-established ecotourism destinations including countries in Central and Latin America, Africa, Asia, and the Pacific. Island destination areas include the Caribbean, Mexico, the Galápagos, and the South Pacific.
4   *Hawaii's dependence on a mass tourism economy.* Ecotourism's contribution to Hawaii's economy will be modest in comparison to Hawaii's mass tourism industry and should be viewed in terms of diversification, not substitution.
5   *Hawaii's popular image.* As an ecotourist destination, Hawaii will have to overcome its sun, sand, and surf reputation and the impression that it is too overdeveloped to appeal to outdoor enthusiasts seeking back-to-nature experiences.
6   *Private-sector investment.* Owing to the relatively low-return/high-risk nature of the ecotourism industry, venture capital and investment support for small business enterprise development are limited while liability costs (i.e. insurance) for landowners remain high.
7   *Public policy considerations.* There is presently a lack of policy which formally addresses ecotourism issues, resulting in inadequate support for the development and maintenance of ecotourism resources.

These barriers nicely encapsulate the constraints that Hawaii will encounter in its planning and delivery of ecotourism in the state. Other regions should be equally wary in the development of their ecotourism product. Substantiating ecotourism through inaccurate research methods is as disruptive as the development of an ecotourism industry on the basis of other related, and less ecologically sensitive, forms of tourism.

As a final note to this section, Diamantis and Ladkin (1999) write that today few tourism and hospitality organisations are proactive when it comes to the environment. They point out that only 19 per cent of hotels surveyed had an environmental policy, and of these only 50 per cent publicised these policies to their guests. Diamantis and Ladkin suggest that better practices will need to be employed if companies are to progress beyond the repositioning of products that appear to be eco-friendly, or applying the three R's in appearing green. They note that the environmental marketing initiatives developed during the 1990s demand better planning and implementation strategies beyond the 'window dressing' approaches outlined above. As described by Peattie (1995), green marketing is the holistic management process responsible for identifying, anticipating and satisfying the requirements of customers and society in a profitable and sustainable way (cited in Diamantis and Ladkin 1999). Green marketing differs from other conventional approaches to marketing in the folllowing four ways:

1   It has an open-ended rather than a long-term perspective.
2   It places greater emphasis on the natural environment.

3   It treats the environment as something that has an intrinsic value over and above its usefulness to society.
4   It focuses on global concerns rather than those of individual communities.

The notion here is that market share can still be accessed and increased, environmental performance of goods and services can be improved, and the integrity of the environment and the people who inhabit these spaces can be maintained.

## Management issues in ecotourism

## Privatisation

The idea behind the partnering of public and private sectors is not a new one. Adams (1983) reports that in the realm of community development, such endeavours came into being during the early 1970s. He further illustrates that a number of trends have emerged that help define the scope of such public–private interactions, which include the fact that:

1   Local governments are becoming increasingly concerned with maximising efficiency through improving management and streamlining service systems.
2   New ways of leveraging existing public resources are being explored.
3   Local governments are beginning to alter their delivery systems to use existing resources in new ways.
4   Cities and counties increasingly are having to consider almost completely devolving certain responsibilities back to citizens.

In general, privatisation, in the sense used here, refers to an arrangement entered into by an employer and an outside firm whereby production or service work which was or could have been done by their own employees and equipment is to be performed by the outside firm (Young 1964). Privatisation is, in most cases, synonymous with the contracting out of services, but emphasises the *private sector's* role in fulfilling or winning these contracts (it is important to note that contracts may also be given to those in the not-for-profit sector). Similar in nature to the aforementioned definition is one by Courchene (cited in Beres 1986), who wrote that privatisation is the private provision of a publicly financed service; while Heald (1984) suggests that privatisation is the strengthening of the market at the expense of the state.

As a result of the increasing pressure on governments to be more accountable with fewer resources, and in the absence of any change in this trend in the foreseeable future, public service providers are caught in the dilemma of how best, in the light of the changing socio-economic conditions, to accomplish traditional goals such as service provision and accessibility. Given these changes, it is safe to say that in all probability there are few communities that have not endeavoured to privatise in one way or another. This assertion is best summed up by Beres (1986: 4), who illustrates that good managers have always looked to the private sector when it did not make sense to employ people on a full-time basis.

In an extensive overview of the topic of recreation policy in Canada from the municipal perspective, Smale and Reid (1995) suggested that the decision to privatise, or not, is really a reflection of the perceptions and interpretations of what is a public good. On the one hand, these authors feel that partnerships with the volunteer and private sectors may be necessary in order to deliver as much as possible to the entire community. On the other hand, they felt that in the debate on privatisation, policy-makers must consider what they value in providing for the community:

privatisation may be interpreted as a public good in today's ideological world which places an emphasis on the *profitability* rather than the *accessibility* of leisure opportunities. If the service can be provided through the market place, then it may very well end up there without great concern for its value as a public good. Consequently, even though the service may become 'successful', it would likely only serve a certain segment of the population, and hence, would no longer be available to the entire community; in other words, the principle of equal opportunity would be compromised.

(Smale and Reid 1995; emphasis in original)

Mintzberg (1996) is a firm believer in the notion that capitalism driven by the private sector is throwing the societies of the Western world out of balance. He would rather see a move to the incorporation of normative control, or values, beliefs, and human dedication in the provision of goods and services (the Japanese have been successful using this approach). He continues by suggesting that 'we all value private goods, but they are worthless without public goods – such as policing and economic policies – to protect them' (p. 117). Mintzberg substantiates his claims by referencing the current preoccupation with privatisation of public services in today's economy. He suggests that in some cases privatisation is warranted, but in most other cases we are too prone to scrutinising what does not belong in government when we should be equally diligent in assessing what does not belong in the realm of private business.

In areas more central to tourism, privatisation has generated much debate in the realm of destination promotion at the state and national levels. Bonham and Mak (1996) provide a good overview of the history of private and public financing of Hawaii as a destination, which they use to put the private/public promotion debate into context:

Faced with growing voter opposition to public financing of tourism promotion, governments are increasingly asking the tourist industry to assume a larger share of the burden of financing destination promotion. However, asking the industry to make greater voluntary contributions will not succeed without some form of government intervention. Because destination promotion is a public good, individual beneficiaries have the incentive not to contribute and instead to free-ride on the contribution of others.

(Bonham and Mak 1996: 9)

From a philosophical standpoint, there needs to be some clarity established in better understanding the role that privatisation could play in developing and less developed countries, and in the mass tourism and the ecotourism industries. For example, Jenkins (1994) has illustrated that privatisation will become more prevalent in the Third World as a pragmatic shift in policy slowly over time. He further acknowledges that such commercialisation would enable the industry to become more sensitive to market trends and forces. The private sector's role is one that is concerned with profitability. When ecotourism is involved, caution must be employed from a marketing perspective for the simple reason that preservation of ecosystems and just social and economic practices may be compromised in an all-out assault on profit. Tisdell (1995) writes that investment in ecotourism takes the form of private firms and public agencies intent on making a profit or surplus from the development of ecotourism. In the case of the public agencies, investment usually occurs in an attempt: (1) to make a profit or surplus from ecotourism to supplement its funds for conservation management; (2) to add to political support for the conservation body, or avert political hostility or lack of political support; and (3) to improve the working conditions and salaries of employees.

Eagles (1995) feels that private reserves have a role to play in the provision of ecotourism services in Canada. Notwithstanding, he feels that parks services (i.e. Ontario Provincial

Parks and Parks Canada) should be flexible enough to do a better job at financing their parks. He cites the fact that such parks will be ill prepared to accommodate visitation in the future as they increase in number with continued decreases in funding. Eagles feels that in order for parks to be more business-like, they will have to look closely at revenue retention, willingness to pay, and fees for service as means by which to balance their books, visitor satisfaction, and resource protection.

In a review of ecotourism, less developed countries, and parks and protected areas, Fowkes and Fowkes (1991) identify four different goods in relationship to the element of privatisation. They feel that both common goods, such as air, rain, and fish (individually consumed, and difficult to control), and collective or jointly consumed goods, such as national health services, should not be privatised. On the other hand, toll goods (e.g. telephone systems) and private goods (e.g. foodstuffs), where consumption can be controlled by the supplier, ought to be in private hands. These authors feel that parks and protected areas, or collective goods, should not be transferred to private interests. It is only when toll goods and private goods (accommodation) are provided in parks that privatisation should be considered as a strong option. In most cases involving support services (e.g. computers), retail activities (e.g. restaurants), tourist buildings, and infrastructure (e.g. roads), private intervention should be considered from the perspective of management, maintenance, control of access, marketing, and construction. Private involvement in aspects of the land or those that inhabit the land (plants and animals) is not appropriate.

## Not-for-profit organisations and NGOs

Depending on the perspective from which one approaches the concept and delivery of ecotourism, there may or may not be a natural linkage between ecotourism and the not-for-profit sector. In many cases the link that does exist between the two (parks and not-for-profits) is based on the vested interest in how such parks should be preserved in the face of competing usage. Although one should not assume that not-for-profit organisations with environmental mandates are not conscious of the value of money in the administration of their operations, their motives can be thought of as being more park- or education- or ecology-centred than profit-centred when compared to the private sector. While not-for-profit agencies are typically bound by the demands of a board of directors whose responsibility is to direct programme and operational policy and decision-making within the agency effectively, earnings that are made beyond operational costs (including salaries) are poured back into the programmes rather than into the pockets of management/employees.

Gardner (1993: 19) illustrates that an environmental non-government organisation (ENGO) is a citizens' group whose activities include efforts for environmental conservation. Generally these coalitions have the following characteristics: (1) voluntary membership in the group; (2) the group is not profit motivated; (3) it is autonomous, in that it makes its own decisions; (4) it is service driven; and (5) it seeks change on behalf of its membership, society, and the environment. However, NGOs are simply defined as 'not government', and can take the form of not-for-profit status or profit status. There appears to be an interesting dilemma evolving on this front. In many cases not-for-profit agencies (e.g. museums, universities) are heavily involved in the delivery of ecotourism programmes *for profit*, which, in theory, is inconsistent with their overall mandates as institutions. Ziffer (1989) has suggested that not-for-profit sector agencies often sponsor ecotourism trips for a number of reasons, including member service, donor trips, a source of funding, and for education and research. Citing Ashton, a former not-for-profit trip provider, Ziffer comments that many not-for-profit agencies are making a significant amount of money from their trips.

In a survey of US-based nature tour companies, Higgins (1996) illustrated that the not-for-profit sector has a firm stake in the delivery of ecotourism opportunities. Eleven

## CASE STUDY 5.1

### *Highlighting the not-for-profit sector: Conservation International*

Conservation International (CI) is a not-for-profit organisation that strives to conserve the earth's biodiversity along with demonstrating that human beings are able to live harmoniously with nature. One of CI's most significant hallmarks is the creation of the Dept-for-Nature Swap, which gives debt-poor nations a chance to make smart conservation choices. CI negotiates to buy their foreign debt in exchange for national conservation initiatives. Native habitats that were in danger of disappearing are now safe havens for thousands of threatened species.

CI's mission in ecotourism is to develop and support ecotourism enterprises that contribute to conservation; and influence the broader tourism industry towards greater ecological sustainability (Conservation International 1997). In order to accomplish these ends, CI is involved in a number of national and regional ecotourism development initiatives around the world in countries such as Bolivia, Brazil, Guatemala, Peru, Botswana, Madagascar, Indonesia, and Papua New Guinea. CI uses a 'capacity-building' approach in these regions to ensure that ecotourism benefits communities and merges with traditional practices and conservation, through the training of local people via ecotourism workshops within the region. Recently, CI has developed an Ecotravel Center designed to provide information on ecotourism destinations, tour operators and lodgings, and relevant publications and information.

tour operators (17 per cent of the sample total) were not-for-profit, serving some 20,215 clients. Forty-two per cent of their client trips were offered in North America, which compared to 7 per cent of the rest of the sample, which were commercial in nature. Also, it was suggested by Higgins that the not-for-profit sector was quite active in securing its share of the market by using 'sophisticated marketing campaigns, including direct mail, advertising in tourism publications, and the development of glossy nature-tour brochures' (Higgins 1996: 16). In other research, Weiler (1993) found that on the basis of a content analysis of 55 tour operators' brochures involving 402 tour descriptions, 336 were private-sector tours and 66 were not-for-profit or university-run tours. Universities are given the status of not-for-profit, yet they sponsored and ran programmes designed to make money.

The jury appears to be still out on the relationship between the not-for-profit agencies in ecotourism. One of the main issues related to the profit and not-for-profit sectors is the quality of the tours that they run. Although no data were found in order to elaborate the point, it is nevertheless an issue of great concern for ecotourists and members of host communities. Given ecotourism's local development mandate one could argue that the not-for-profit agencies (all things considered equal) would be the more legitimate tour operators. This is supported by Ziffer (1989), who feels that the not-for-profit sector is a small but growing segment of the ecotourism industry with plenty of potential in establishing appropriate ecotourism development models for the future (see Backman *et al.*, 2001, for a discussion on the blurring of for-profit and not-for-profit roles in the development of ecotourism programmes).

## CASE STUDY 5.2

### *Ecotourism and Monarch butterflies*

One of North America's great yearly migrations is undertaken by the Monarch butterfly which, in the millions, flies upwards of 8,000 km from Canada/US to one of 15 small overwintering sites spread over only 50 acres in the Transverse Neovolcanic belt of Mexico. In this region, which is marked by fir trees, sufficient moisture, wind protection, and nectar, is the habitat they need to survive. Here they will stay for about five months before venturing off north. The most astonishing aspect of this migration is that no single butterfly ever makes the entire round-trip. As suggested by Halpern (1998), three to four generations separate those that came from Mexico from those that return. Halpern says that a Monarch born in August in New York flies all the way to Mexico to overwinter, and begins the northward journey in March, laying eggs along the coast from Texas to Florida before dying. Butterflies born in Florida, for example, continue north breeding and laying eggs along the way, as do their offspring. By August, another butterfly, four generations removed from the one born in New York the previous summer, will metamorphosise in New York and repeat the steps. In doing so it goes to a place, a small parcel of land in Mexico, that it has never been to.

While the natural history associated with this story is a great mystery of science, the human component is rather less spectacular. Barkin (2000) writes that the creation of a special biosphere reserve to protect the Monarch from human encroachment (he says as a result of a 1976 article in *National Geographic Magazine* by Urquhart which detailed the journey of the Monarch), had many catastrophic implications for local communities. The NGO responsible for establishing the reserve, through the World Wildlife Fund, unfortunately did not undertake any consultation with the local community. In establishing protection for the butterflies, no compensation was afforded to the locals who, without the ability to harvest resources from the protected areas, were unable to earn a living. And while some facilities were constructed to cater to tourists, in no way are they able to accommodate the needs of 250,000 (1998–99 figures) – representing a tenfold increase from a decade earlier.

Not only did the local people lose the ability to control their lands, the benefits of ecotourism were only accrued by a relatively small percentage of the inhabitants of the region. In many cases, the shortage of jobs forced some people to leave their families behind in order to search for other low-paying jobs as far away as Mexico City (three hours distant) to support their families. Still others, who traditionally logged the forests of the region, were forced through necessity to go back to this way of life illegally. Aridjis (2000) and Halpern (1998) observe that even those in charge of reserves have been involved in illegal logging, due to the corruption which has developed through the social and economic dislocation which has become common in the region. Some individuals have resorted to putting large gashes in the trees so they eventually die, in order to take them 'legally'. Added to this is the feeling that some of the reserves are in danger, as lucrative logging concessions are awarded in exchange for political patronage. This pits those who depend on the reserve for tourism directly against those who would earn a living through logging.

While some have suggested that ecotourism is one of the few ways to rectify the problem (e.g., through developing small guest houses and having children and women sell souvenirs), others are not so optimistic. Halpern notes that such a scheme is unrealistic because in the absence of sound planning there is a lack of infrastructure to support a growing number of tourists, it does not take into account the impact from these tourist numbers (trail erosion, garbage, water pollution, and so on), and there are no regulations in place to limit the number of tourists in the same way that logging was regulated in the past. There is also the issue of environmental education that appears to be seriously lacking in guides and tourists. Halpern notes that an American 'naturalist' who wanted to go past a barbed-wire fence to get closer to the butterflies, paid a guide US$8 to lift the fence. In getting closer to the butterflies, the tourist trampled upon a newly seeded part of the reserve, killing some of the seedlings in the process.

From the perspective of Barkin, what is needed in attempting to ameliorate the problem is an approach to break the cycle of impoverishment and environmental degradation. He says that ecotourism can provide opportunities for local people, but only as a complementary activity to other land uses. He proposes the following:

● Government must recognise locals' right to be in control and benefit from the Monarch butterfly, through training, technical assistance, and resources.
● Need mechanisms to ensure the traditional system of food production is maintained.
● To consider the relevance of sustainable tourism practices for the region, although it is unlikely that the Mexican government will recognise this form of tourism for rural development.
● To explore the potential to harness geothermal energy for agro-industrial, recreational and medicinal purposes (spas).
● To address the seasonality problem through the promotion of cultural tourism and other aspects of the region's biodiversity.
● To spread the benefits of tourism to many, instead of resting it in the hands of a few.
● To move away from the current tourism policies which are environmentally destructive, contributing to social polarisation with a great deal of leakage.
● Attempt to integrate tourism into regional development for a more balanced economy.

This situation is quite similar to the problem in Tres Guarantias, Mexico (see p. 156), where the imposition of ecotourism from a top-down perspective has proven disastrous to local people. Again, the aspect of control emerges as a critical element in the development of an ethical ecotourism industry. Communities are maintained in an equilibrium state before the imposition of tourism (tourism which will benefit a few). Time, education, control, and the intrinsic nature of involvement need to be factored into the formula for any degree of success within such an enterprise.

## Conclusion

This chapter has taken a broad-based look at how economics, marketing, and a series of other management issues related to ecotourism. Ecotourism has blossomed into a rich array of programmes and opportunities in the public, private and not-for-profit sectors. Often a relationship must exist between public agencies (those who manage parks and protected areas) and private operators (those who run programmes in these public spaces). Good coordination must exist between these stakeholders such that appropriate values (economic and emotional) are attached to the ecotourism experience. A significant finding of the chapter is that marketing must better reflect the needs not only of the consumer but also of the industry, which by many accounts is being diluted by ecolabelling, or the misrepresentation of ecotourism by operators who at best are only peripherally ecotourism oriented in their product offerings. Finally, the chapter emphasised the need to examine how privatisation, not-for-profit agencies and NGOs affect the delivery of ecotourism products.

## Summary questions

1   How do alternative forms of tourism, like ecotourism, act as better options in helping to diminish leakages in an economy?

2   Using the example of the Monteverde Cloud Forest Reserve in Costa Rica, how has a differential fee structure been utilised in such an area?

3   What do the travel cost method and contingent valuation method allow economists to do? Why are these important?

4   Some regions boast that 'because wildlife pays, wildlife stays'. What does this mean, and how is it that ecotourism might contribute to the preservation of species?

5   As not-for-profit agencies are philosophically different than private/commercial agencies, how is it that ecotourism might benefit from the philosophies of these not-for-profit agencies in regard to ethical conduct and programme excellence?

## 6

**From policy to professionalism**

Buckley and Pannell (1990) write that environmental impacts might be minimised through the use of a combination of factors including planning and regulation, incentives to encourage or discourage particular activities, hardening of sites receiving high levels of use, and through the use of education both on- and off-site. Such policy-related actions are increasingly required, according to Goodwin (1995), to ensure that tourism development is consistent with the needs of both local people and the environment. Tourism and ecotourism policy encompasses a broad spectrum of concerns related to the implementation of tourism programmes around the world, including social, ecological and economic relationships; it also considers how tourism affects or is affected by tourists, local people, operators, government, and so on. In addition, much discussion has surrounded the notion of regulation as a means by which to provide a stronger element of control in how ecotourism products are shaped and implemented. Regulation, therefore, is also an integral part of this chapter. The final part of the chapter deals with certification and accreditation in the ecotourism industry, as well as guides and interpretation. It is suggested that the industry is committing itself further to these programmes as a means by which to offer a higher standard of quality and professionalism in the field.

## Tourism policy

In a simplistic sense, tourism policy is the identification of a series of goals and objectives which help an agency – usually a governmental one – in the process of planning the tourism industry. According to Akehurst (1992), however, policy development is much more detailed. It is further defined as:

> a strategy for the development of the tourism sector . . . that establishes objectives and guidelines as a basis for what needs to be done. This means identifying and agreeing objectives; establishing priorities; placing in a Community context the roles of national governments, national tourist organizations, local governments and private-sector businesses; establishing possible co-ordination and implementation of agreed programmes to solve identified problems, with monitoring and evaluation of these programmes.
>
> (Akehurst 1992: 217)

In this sense it is the coordination of many organisations and agencies involved in the provision of tourism services, and the planning, development and management of these groups.

It is felt that all countries should endeavour to create tourism policy to guide their planning, management and development of tourism consistently throughout a region and as a means by which to use resources in a wise and efficient manner (Jenkins 1991). The sector most responsible for the impetus to create policy is government, in either a passive (e.g. legislation is introduced but is not intended to discriminate in favour of the tourism industry) or an active manner, where government takes action to discriminate in favour of the tourism sector managerially (through the creation of objectives and legislative support) and/or developmentally (in the establishment and operation of tourism facilities). In a similar context Lickorish (1991) suggests that governments have given low priority to the establishment of policy in tourism, particularly in the developed world (the first policy reports were established by the European Community in 1986). The reason for this is that governments have traditionally viewed tourism development as lying within the realm of the private sector (Lickorish 1991; Pearce 1989). If centralised action has not been taken within a country, often regions or localities have taken responsibility – but, according to Lickorish, with uneven results. In addition, policy-related problems are compounded by the fact that government policies related to tourism and the environment can be quite divergent. Ministries of agriculture, the environment, tourism and industry, and transport may all vary in their approach to environmental policy.

Coccossis (1996) feels that there has not been a good understanding between the complex nature of the environment and tourism (e.g. the synergistic effects of tourism), coupled with a high degree of administrative fragmentation. In addition, he writes that environmental conservation was perceived as a threat to social and economic development; a threat which has only recently subsided as a result of our efforts towards the clarification of sustainable development. Without due regard to appropriate development guidelines, the integrity of public and private lands within communities has often been at risk owing to the sacrificing of socio-ecological values at the expense of profit. A key point in Coccossis's argument is that environmental policies now are characterised by a more holistic, ecosystem based approach that relies less on specific issues and more on a regional integrated perspective. This wider approach involves the following (pp. 10–11):

1   linking development policy with environmental management. As a first step, the review of projects, plans and programmes from an environmental point of view should be instituted;
2   regional-level environmental management schemes which would provide a framework for guiding local environmental management programmes;
3   integration of tourism development and environmental management policies at the local, regional, and national level; and
4   increasing local capacity to cope with environmental issues, particularly in rapidly developing tourist destination areas.

Priority areas identified by Coccossis (1996) in Europe for sustainable tourism development, utilising the aforementioned concepts, include coastal areas, rural areas, built environment and urban areas, and islands. One of the best examples of policy related to hotel development was the implementation of measures to control the height of hotel development in Bali, Indonesia, and other places around the world in the 1980s. Hotels could not be built higher than the tallest palm trees to avoid the unsightly nature of these developments. Hence, the development would fit into the landscape rather than dominate it.

In support of what Coccossis illustrates above, Fayos-Solá (1996) writes that tourism policy has started to change in line with the evolving nature of the tourism industry. Consistent with the paradigmatic shift in the 1980s away from mass tourism to alternative forms of tourism, the industry has given way to globalisation of markets, supersegmentation, new technologies, and an increased sense of social and ecological responsibility. Competitiveness is a key function of the tourism industry and it is based on quality and efficiency rather than the quantity, which has been the *modus operandi* of mass tourism.

Conversely, tourism policy objectives have undergone three distinct generations, according to Fayos-Solá (1996). The first generation was based on large numbers of visitors, the maximisation of revenues from tourism, and the creation of numbers of jobs, indicative of the mass tourism paradigm. The second generation evolved from the economic difficulties of the 1970s, including growth–recession fluctuations. In this stage, social, economic and ecological impacts are better understood, while the economic and legal objectives are redefined in tourism's contribution to societal well-being. The final generation shows competitiveness as a key function of the vitality of the tourism industry. In addition, both total quality management and the partnership between public, private and not-for-profit agencies is underscored, leading to a more symbiotic relationship between sectors. As such, the politics of policy-making for the twenty-first century will go beyond the conventional discussions related to marketing, promotion, tax incentives, accommodation and transportation to more holistic-centred issues related to the environment, social impacts, issues of rational and equitable access (who gets what, when, and how), and the international regulation of health and safety issues (Richter 1991). Richter further illustrates that the political interest in policy is coming of age and that the democratisation of the electorate may prove to be formidable proponents in shaping key tourism decisions.

In a different line of thinking from the policy issues outlined above, Hjalager (1996) argues that innovation must be the key to future policies and regulation designed to limit tourism's impact on the resource base. She feels that policy, when it is considered, is hampered first by the fact that it is governed by ideology and personal sympathies rather than the need for balance, and second by the fact that cost is unlikely to be considered

**Table 6.1** *Innovation's place in tourism policy and regulation*

| Instruments | How the environmental effect is achieved | Influence on innovation |
| --- | --- | --- |
| Emission standards. Inspection in order to control | Industrial 'end of pipe' norms expressed according to carrying capacities. Development permissions with environmental standards | Control may raise standards, severe control may lead to the development of new techniques. No evidence that tourism will take the innovation lead |
| Compulsory use of specific energy resources or specific technologies (e.g. district heating or waste treatment systems) | Obtaining economies of scale in environmental management systems | Typically, innovation will not take place in tourist industries, but with specialised plants and suppliers |
| Zoning in order to limit or control opportunities for development Control of land use | Limiting the developer's activities according to carrying capacities or environmental objectives | Probably none |
| Zoning of tourists' access to vulnerable resources or areas | Visitor control systems and regulation of tourism by volumes Compulsory methods or training, motivation | Opportunities for development of IT instruments combined with communication/interpretation instruments |

*Source*: Hjalager (1996)

in implementation. Her research touches on three policy control mechanisms: the market, bureaucracies and clans (industry in cooperation with other stakeholders). For example, Hjalager writes that bureaucracies regulate tourism in basically two fashions: (1) by directing tourists, locals, and the industry through legislation, and (2) through the provision of tourism infrastructure (see Table 6.1). Central to Hjalager's thesis is the notion that regions must be open to alternative modes of regulation in addressing significant and challenging environmental issues.

## Ecotourism and policy

In many countries ecotourism is presently at the policy consideration and initiation stage, with the recognition that further political and socio-economic coordination must exist for it to proceed. According to Weaver (2001), this means that national or regional ecotourism plans are more meaningful if related to and informed by a broader national tourism policy, which itself is a manifestation of an overall government policy. And as suggested by Sofield and Li (2003), plans will fail to be implemented if they do not fit the existing policies of government. Compounding this issue is the notion that even with policies in place, at times governmental agencies responsible for upholding policies have been found to work in isolation from one another, creating a great deal of counterproductivity (Jenkins and Wearing 2003).

Policy development has only started to be initiated as a result of the lack of consensus as to what constitutes appropriate ecotourism development and activities. This has been the case in Hawaii, as identified by the Center for Tourism Policy Studies (1994). This group

---

### CASE STUDY 6.1

#### *The regulation of whale-watching in Canada*

The impact of ecotourism on the beluga whale population of the St Lawrence River, Canada, has prompted Blane and Jackson (1994) to call for stricter policies in the control of such vessels. The authors found that the high number of trips/vessels in the area of the whales was affecting their behaviour. They recommended that the number of operators should be restricted, that low speed limits should be spatially increased and enforced, and that boats should be required to travel around beluga pods in an appropriate fashion. Further, the authors suggested that policing should be accomplished by the Canadian government and informally by peers (i.e. operators). Such regulations have been addressed on the west coast of Canada by whalewatching operators, government, and other interested groups, and a set of guidelines for the industry to follow has been developed. A recent draft of a document by Whale Watching Operators North West advocates the following: (1) stay approximately 100 metres away from the whales; (2) operators should be familiar with the Canadian Federal Fisheries Act; (3) slow down half a mile before arriving at the whales; (4) whales should be approached slowly at any time; (5) when approaching whales from the front, STOP and allow whales to come to you; (6) when approaching from behind, move to the outside of the nearest whales or boats, parallel to the direction of travel at a similar or slower speed; (7) when travelling with the whales, do not alter speed or change course abruptly; (8) never make high-speed runs through the middle of a group of whales or boats; (10) avoid approaching whales that are feeding; (11) try to work with the whales in rotation; and (12) do not reposition your vessel by the leap-frog method.

has identified a number of policy recommendations to address in developing Hawaii's ecotourism industry, including: (1) a state ecotourism policy and ecotourism development plan; (2) interagency coordination and cooperation with the private sector; (3) increased funding support through direct and indirect means; and (4) active community participation in planning and decision-making. These recommendations are further broken down into integrated planning, public- and private-sector roles, research, land use, conservation, preservation, funding, marketing, operator concerns, socio-cultural concerns, regulation, monitoring, and education and training. As a prime US destination, Hawaii is realising that apart from its extensive mass tourism industry, there is a definite market for ecotourists based on interests in whale-watching and the rainforest. Although rough estimates indicate that ecotourism represents about 5 per cent of Hawaiian tourism revenue, there is the feeling that ecotourism is on the rise in this state (Marinelli 1997), which is good news given the negligible tourism growth in Hawaii over the past decade.

Much like the Center for Tourism Policy Studies cited above, Liu (1994) suggests that government policy is one of the keys to supporting the development of ecotourism. Her work on behalf of the American Affiliated Pacific Islands discusses the importance of government policy as a means by which to regulate and monitor the industry in a way that balances protection and restriction without hindering the individual operator. This means a balance between development and conservation, supply versus demand, benefits versus costs, and people versus the environment. In addition, Liu feels that government must play a leadership role in providing the necessary financing, management skills, and knowledge so the private sector can operate smoothly and efficiently. This means government must (Liu 1994: 8, 9):

1 facilitate efficient private-sector activity by minimising market interference and relying on competition as a means of control;
2 ensure a sound macro-economic environment;
3 guarantee law and order, and the just settlement of disputes;
4 ensure the provision of appropriate infrastructures;
5 ensure the development of human resources;
6 protect the public interest without obstructing private-sector activity with too many regulations;
7 promote private-sector activity by not competing in the business arena with private enterprise; and
8 acknowledge the role of small business entrepreneurs and facilitate their activities.

Liu (1994) has developed a public-sector guide for the implementation of ecotourism, as set out in Table 6.2. These are the points that should be considered for the purpose of implementing appropriate ecotourism policy.

Increasingly, planners and developers are realising that policy must involve the many stakeholders who stand to be impacted by ecotourism development, as suggested at the outset of this chapter. According to Ceballos-Lascuráin (1996a: 85–91), some of these groups include protected-area personnel, local communities, the tourism industry, NGOs, financial institutions, consumers and national ecotourism councils. Although this is an essential aspect of policy development from the perspective of inclusion, Fennell *et al.* (2001) note that there will be problems in achieving compromise between so many stakeholders in the tourism sector within a particular planning jurisdiction. This is particularly true of ecotourism, which must balance use and preservation, as suggested by Liu (1994) earlier.

One of the finest examples of policy development involving stakeholder involvement in ecotourism can be found in the Australian National Ecotourism Strategy (Commonwealth

**Table 6.2  *Policy implementation framework***

*Development objectives.* Establish economic, ecological, and socio-cultural objectives in consultation with local communities; designate specific areas for ecotourism development.

*Inventories.* Survey and analyse the region's ecology, history, culture, economy, resources, land use, and tenure; inventory and evaluate existing and potential ecotourist attractions, activities, accommodation, facilities, and transportation; construct or consolidate development policies and plans, especially tourism master plans.

*Infrastructure and facilities.* Provide the appropriate infrastructure and facilities, avoiding a reliance on foreign capital; establish means to assist the private sector in developing ecotourism enterprises in line with ecological and cultural standards.

*Market.* Analyse present and future domestic and international ecotourism markets and establish marketing goals; know and understand the market in achieving goals; assist the private sector in its development of marketing strategies.

*Carrying capacity.* Strive to understand the social and ecological limits of use of an area through appropriate management and research; establish social and ecological indicators of use and impact; implement an appropriate pre-formed planning and management framework.

*Development.* Establish a development policy giving consideration to balanced economic, ecological, and social factors; form a development plan on the basis of attractions, transportation, and ecotourism regions; assist developers to plan and build ecologically.

*Economic.* Consider ways to enhance economic benefits; conduct present and future economic analyses; ensure that profits are made, locals benefit, and public revenues are self-sustaining.

*Environment.* Consistently evaluate the impact of ecotourism on the resource base; link ecotourism with other resource conservation measures (e.g. parks and protected areas).

*Culture.* Evaluate the socio-cultural impact of ecotourism, prevent negative impacts, and reinforce positive outcomes; empower local people to become decision-makers; conduct a social audit of social impacts.

*Standards.* Apply development and design standards to facilities and accommodation; facilitate the adherence to standards by providing financial or tax incentives and access to specialists.

*Human resources.* Promote job creation and entrepreneurship; establish community awareness programmes; provide adequate education and training for local people.

*Organisation.* Establish a working relationship between public, private and not-for-profit organisations.

*Regulations and monitoring.* Establish legislation/regulations to promote ecotourism development, through support for tourism organisations, tour operators, accommodation; establish facility standards.

*Data system and implementation.* Establish an integrated ecotourism data system for continuous operation that provides research and marketing information; identify ecotourism implementation techniques; and collaborate with private industry and educational institutions in implementation.

*Source*: Liu (1994)

Department of Tourism 1994). In this document an integrated approach to the development of ecotourism has been adopted with the belief that development and management of ecotourism is fundamental to optimising the benefits it offers. This strategy is truly a national document in that it integrates the collective opinions of Australians through the implementation of a series of public consultation workshops involving government, industry, conservation groups, educational institutions and community groups. (The stakeholder approach used in this strategy has been outlined by Boo, 1992, who suggests that there are several stakeholder groups that must be involved in the development of ecotourism initiatives.) The intent of the strategy is summarised as follows: 'to provide broad direction for the future of ecotourism by identifying priority issues for its sustainable development and recommending approaches for addressing these issues' (Commonwealth Department of Tourism 1994: 6). The bulk of the strategy concentrates on a variety of issues, objectives and actions which are essentially steps towards putting policy and planning initiatives into practice. The 12 objectives of the strategy are set out in Table 6.3.

It is the ultimate aim of these directives to move Australia to create an ecologically and culturally sustainable ecotourism industry in line with the ideals of sustainable development. The development of this document was instrumental in demonstrating to the rest of the world the advanced state of ecotourism in Australia, and it is a model which other governments can follow. For example, the Ecotourism Task Force, in association with Tourism Saskatchewan (the government–private industry consortium that leads tourism development in this Canadian province), has developed a similar document which will be the basis of the ecotourism industry for the future. Recommendations (among others) include the development of an accreditation process, the development of cooperative relationships with other land-users, and the construction of sustainable ecolodge facilities using state-of-the-art technologies and modelling.

As regions and countries make the decision to plan and develop an ecotourism industry, Ceballos-Lascuráin (1996a) advocates following a basic planning process such as the one identified below. Policy may naturally evolve out of first identifying many of the catalysts and constraints to the development of the industry. In the case of ecotourism, this process must proceed out of government's general development policy and strategy (if such a strategy exists).

1   *Study preparation.* Includes the assessment of the type of planning required and the preparation of terms of reference.
2   *Determination of objectives.* Such objectives must reflect the national or regional government's general ecotourism policy/strategy, and include development priorities, temporal considerations, heritage, marketing and annual growth.
3   *Survey.* A complete evaluation and inventory of existing resources must be made, especially related to the attraction base. The ultimate aim of this inventory is to link attractions to various market segments and forms of development.
4   *Analysis and synthesis.* This step involves studying the historical background of tourism in the region, analysing constraints to development, legal and risk management considerations, financing, tax incentives, protection of cultural and natural features, and other economic-related variables (e.g. contribution to GNP, and complementarity with other sectors of the economy).
5   *Policy and plan formulation.* From an analysis of the synthesis, policies must be structured to reflect the economic, social and ecological needs of the region. Alternative policies should be developed to assess how each fits with the country's overall development policy, from which final policies are derived in the areas of: infrastructure, human resources, transportation, intersectoral coordination, establishment of councils and committees, tax incentives and subsidies, and the creation of tourism programmes.

**Table 6.3** *Australian National Ecotourism Strategy objectives*

| Strategy component | Objective |
| --- | --- |
| Ecological sustainability | Facilitate the application of ecologically sustainable principles and practices across the tourism industry |
| Integrated regional planning | Develop a strategic approach to integrated regional planning based on ecologically sustainable principles and practices and incorporating ecotourism |
| Natural resource management | Encourage a complementary and compatible approach between ecotourism activities and conservation in natural resource management |
| Regulation | Encourage industry self-regulation of ecotourism through the development and implementation of appropriate industry standards and accreditation |
| Infrastructure | Where appropriate, support the design and use of carefully sited and constructed infrastructure to minimise visitor impacts on natural resources and to provide for environmental education consistent with bioregional planning objectives |
| Impact monitoring | Undertake further study of the impacts of ecotourism to improve the information base for planning and decision-making |
| Marketing | Encourage and promote the ethical delivery of ecotourism products to meet visitor expectations and match levels of supply and demand |
| Industry standards/ accreditation | Facilitate the establishment of high-quality industry standards and a national accreditation system for ecotourism |
| Education | Improve the level and delivery of ecotourism education for all target groups |
| Involve indigenous people | Enhance opportunities for self-determination, self-management and economic self-sufficiency in ecotourism for Aboriginals |
| Viability | Examine the business needs of operators and develop ways in which viability can be improved, either individually or through collective ventures |
| Equity considerations | Seek to ensure that opportunities for access to ecotourism experiences are equitable and that ecotourism activities benefit host communities and contribute to natural resource management and conservation |

*Source*: Commonwealth Department of Tourism (1994)

6 *Recommendations*. The result is a plan that indicates attractions, tourism development areas, transportation linkages, tour routes, and design and facility standards. Also, recommendations are made for implementation, zoning, land use plans for the future, economic benefits, education and training, ecological and social impacts, private industry incentives and legislation.

7 *Implementation and monitoring*. Prior to implementation, the policies and overall plan should be reviewed and ratified legally. Formal review periods should be established and committees or corporations should be developed to help implement or guide the implementation of the various developments.

The prime mandate of many state and provincial tourism authorities is to bolster tourism and ecotourism in their various regions through the development of new product strategies to capture a larger travel market. Given the marketing premise under which many of these regional bodies operate (they are designed primarily to generate profits through marketing and incentives), the social and ecological elements associated with sustainable tourism and ecotourism often take a back seat. However, with increasing frequency subcommittees are being created within these agencies to deal specifically with ecotourism and sustainable development. For example, in 1990 the California legislature's Senate Select Committee on Tourism and Aviation convened to address the many problems facing the California tourist industry, including energy shortages, the inhospitality of cities, and development overkill in resorts. From this committee's study of these and other dysfunctions, it was suggested that a strong ecotourism ethic be created that could be advantageous economically, socially, and ecologically, for the purpose of generating more positive tourism experiences in the state. In part because of some of these early examples of policy development at the regional level, in 1997 I sent out an e-mail to all North American provincial and state governments requesting information on ecotourism policy and accreditation (over sixty offices). Although the response was rather limited, the overwhelming consensus among regional bodies was that policy and accreditation are 'on the table', meaning that while they have been thought about, little has been done in the way of legislation.

In some cases regions are in the process of stating recommendations to government on how to proceed in the ecotourism sector. For example, Florida recently developed a report, based on the deliberations of an advisory committee, on how to protect and plan for its heritage and commercial assets (Ecotourism/Heritage Tourism Advisory Committee 1997), with the mission of developing 'a blueprint that identifies goals, strategies and recommendations needed to create a statewide, regionally based plan to effectively protect and promote the natural, coastal, historical, and cultural assets of Florida, and to link these to commercial tourism in Florida' (p. A–1). The plan focuses on forming strategic relationships with agencies in the community, developing inventories of sites, protection of the environment, education and marketing.

In concluding this section, ecotourism policy may be viewed as being whatever governments choose to do or not to do with respect to ecotourism (see Hall and Jenkins, 1995). This becomes a relevant statement when one considers that, as these authors note, this may involve action, inaction, decision, non-decisions, choice and process. This means that ecotourism policy should be for non-development in the same way it should be for development, as long as the integrity of people and environment is safeguarded in an equitable manner (Fennell and Dowling 2003).

## CASE STUDY 6.2

### *Ecotourism in Brazil*

In the past tourism in Brazil has been defined by many of its coastal regions such as Rio de Janeiro, where the physical nature of the city and its cultural flair have long been admired. More recently, Brazil has also come to be known for its ecotourism potential which rests on a level of biodiversity which is perhaps unmatched in the terrestrial ecosystems of the world. This biodiversity is found in seven main biomes, including the Amazon, sparse forest, savannah, Atlantic forest, pantanel (marshland), southern fields and the coastal zone. Perhaps the best known of these biomes is the Amazon, which is the world's largest forest but also one of the most devastated landscapes. The area itself is 1.6 million square kilometres, covering 18 per cent of the nation, and accounting for 20 per cent of the world's fresh water. However, since the 1970s, ranchers, miners and farmers have destroyed about one-tenth of the forest area (about the size of France). The principal reason for the extent of the deforestation is the need for wood products, but also for subsistence. Because nutrients are so quickly absorbed by plant life, the soil is devoid of the properties which would sustain long-term farming. The result is that families must continue to practice slash-and-burn agriculture to live. Although these practices are frowned upon by the international community, the sheer size of the Amazon forest makes it virtually impossible to enforce guidelines for proper agriculture. The Amazon region can learn from the situation in Brazil's Atlantic forest region, where less than 7 per cent of this ecosystem remains. According to LaFranchi (1998), this region is being deforested at the rate of a soccer field every five minutes, which, at that rate, will eliminate the forest in 50 years. LaFranchi suggests that the new initiatives in ecotourism, sponsored by Conservation International, are the region's best hope to slow the rate of deforestation. This means convincing local people of the value of the forest in an untrammelled state. New initiatives such as canopy walkways have illustrated the power of tourism as a sustainable form of development.

Consequently, ecotourism has been embraced throughout Brazil as a logical alternative to other more consumptive forms of land use like forestry and mining. While there is evidence to suggest that nature-based tourism began in Brazil in 1960, the growth has been characterised as steady, not spectacular. A good indicator of the infancy of the ecotourism industry in the Brazilian Amazon is the fact that most US tourists visit the Amazon in Peru. In 2000, the jungle lodges in the Amazon attracted about 20,000 Brazilian and foreign visitors, which to some is only the tip of the iceberg (Jordan 2001). One of the main constraints, as identified by ecotour operators in the Amazon, is the fact that distances are large and time-consuming; therefore travel is expensive. However, travel is not just expensive in this part of the country – it is costly throughout.

The commitment which Brazil has for ecotourism is represented in the development of a code of ethics for ecotourism in 1996, as observed by Weaver and Schluter (2001), and before this, the publication of an ecotourism policy document which was completed by EMBRATUR (Brazil's tourism institute) in 1994. The following is an example of one of the many 'actions' emphasised in Brazil's Guidelines for a National Ecotourism Policy (Grupo de Trabalho Interministerial MICT/MMA 1994):

*Action 1: Regulating ecotourism*

Objective:

Provide the ecotourism segment with its own legal structure, which is in harmony with federal, state and municipal spheres, and follows adequate parameters and criteria.

Strategies:

- Develop the necessary legal mechanisms for ecotourism development.
- Establish legal control and enforcement mechanisms.
- Identify and evaluate the current legal mechanisms that apply to ecotourism, at all levels of government.
- Propose regulations that are compatible with local reality to environmental and tourism entities, and states and municipalities.
- Define inter- and intra-institutional responsibilities for ecotourism development.
- Establish criteria for prioritising areas with ecotourism potential.

The extent to which these principles filter through the ecotourism industry in Brazil is subject to debate. The presence of many international non-governmental organisations such as the World Wildlife Fund and Conservation International will no doubt help. However, what will also help is a commitment to equity within the industry, and avoidance of what might be referred to as a form of political and economic dominance, which is quite typical in lesser developed countries, that allows just the very few elite to benefit from an enterprise while many others do not. Regional and national ecotourism organisations, tabled by many stakeholders, will also help diffuse information about responsible ecotourism more quickly and effectively. In the eyes of Wallace and Pierce (1996) this job might be a difficult one, based on the results of their study on ecotourism in the ecolodges of the Amazon region. They found that although ecolodges provide employment for local people, they were not adhering to principles of ecotourism. No programmes were developed to protect habitat, guides were not properly trained, and there was a lack of trails and other programme support mechanisms. They further discovered that many of the lodge owners do not even know about the principles of ecotourism, so do not know how to augment their operations as such.

Recently Brazil has been able to leverage over US$200 million from the Inter-American development bank in Washington to develop tourism in the nine states that lie within the rainforest. Beyond this, the government has reportedly spent millions of dollars to develop tourism centred on nature, says Mirabella (1998), who writes that ecotourism has already created hundreds of jobs and put money in the pockets of local people, and forced Brazilians to look at the forest in a different light. The importance of getting local people on the side of this venture cannot be underestimated. In a story on the greening of the Amazon reported in *The Toronto Star*, Romero (2000) says that any conservation project in the region will be useless unless it puts money in the pockets of local people. People simply must be able to feed their families. Romero also notes that small-scale sustainable projects might be the best way for the people of the Amazon to gain control of their economic destiny. This includes tourism, and any other enterprise that allows people to harvest the rainforest in a sustainable manner.

A case in point is the community of Silves, Amazon, where residents who used to sustain themselves through fishing have been forced to consider ecotourism. Fishing, which used to be the mainstay of the residents of Silves, has declined due to population growth,

*continued*

improved fishing technologies, and the destruction of habitat. In looking to the future, local people developed the Silves Association for Environmental and Cultural Preservation to manage the fish reserves. This group has been able to declare specific lakes off limits to fishing, and they have created natural hatcheries as well as developed an ecotourism programme that funds ranger patrols, habitat restoration and education of community members. The development of a simple ecolodge provides comfortable accommodation, local foods and services, with trained guides taking tourists out on the water in canoes and into the forests for hikes.

## Regulation

According to Metelka (1990), regulation refers to the efforts of an agency (government, international or trade) that has been given the authority to regulate the actions of businesses which fall within its jurisdiction. Historically, this has meant that government has seen fit to dictate the actions of, for example, the airline industry in terms of appropriate policies related to safety. The fear, as always, is that if left unchecked tourism will contribute to a diverse range of social and ecological impacts. More recently, the influence of alternative and sustainable forms of tourism has rekindled discussions related to the need to ensure that service and the provision of tourism products stem from appropriate business practices.

The belief that the tourism industry can operate without due regard to its short and long, and direct and indirect impacts is being challenged as more and more there is the recognition that business must be held accountable for its actions. Goodall (1994) provides an excellent overview of regulation through the process of environmental auditing in the context of the British tourism industry. Citing the European Commission, Goodall writes that environmental auditing is essentially a management tool designed to evaluate how well organisations (their management and equipment) are performing towards the safeguarding of the environment through various environmental practices and policies. Goodall (1994) suggests that this environmental tool has technical (e.g. noise reduction technology), legislative (e.g. the 'polluter pays' principle), and business (e.g. the costs and benefits associated with the will to undertake environmental audits) dimensions, which act as both catalysts and constraints to their implementation. In Britain, environmental auditing is voluntary and thus the responsibility lies within the agency, hence it is not likely to happen owing to the competitive, fragmented, small-scale nature of the industry. Consequently, few tourism agencies have formal environmental policies and management systems, and 'Unless required to by law, such firms will only adopt best environmental practice where there is an incontrovertible commercial advantage' (ibid.: 663). In summary, Table 6.4 outlines environmental policy and environmental auditing considerations, as suggested by Goodall (ibid.: 658).

In reference to the UK outgoing tourism industry, Forsyth (1995) discusses the link between sustainability and regulation. His findings suggest that tourism businesses would be in favour of sustainability; however, he feels that because of the fragmented nature of the industry itself (i.e. hotels, tour operators, travel agents, carriers, associations), different industry sectors have separate aims and varied levels of power, and therefore may be unable to implement all forms of sustainable tourism consistently. Another main conclusion of his study was that host governments should be involved in the business of sustainable tourism (through regulation), not necessarily tourism businesses. The implication is that business practice occurs under the umbrella of regional and/or national policy schemes. The problem

**Table 6.4** *Summary of policy and environmental audit objectives*

| Environmental policy objectives for firms | Key aims of an environmental audit |
|---|---|
| 1  Compliance with relevant environmental legislation and indication of a willingness to develop reasonable and workable regulations. | 1  To judge whether the tourism firm's environmental management system is performing satisfactorily. |
| 2  Reduction or elimination of any negative environmental impacts of current activities and avoidance of any negative impacts from proposed developments. | 2  To verify compliance with relevant environmental legislation. |
| 3  Development of environmentally friendly products. | 3  To verify compliance with the firm's stated policy. |
| 4  Sustainable use of resources which include increasing efficiency of resource use, minimising waste, the use of environmentally benign inputs and equipment, and the safe disposal of wastes. | 4  To minimise human exposure to risk from the environment and to ensure appropriate health and safety provision. |
|  | 5  To identify and assess the firm's risk resulting from environmental failure of its activities. |
|  | 6  To assess the impact on the environment, both local and global, of its plant, processes and products. |
| 5  The fostering among employees, and also customers and communities in which the firm functions, of an understanding of environmental issues. | 7  To advise the firm on any environmental improvements it could make. |
|  | 8  To review the firm's internal procedures to help it to achieve its environmental objectives. |

*Source*: Goodall (1994)

with such an approach is that such policies would probably be flexible to accommodate demand of the domestic tourism product and flexible in terms of accommodating multi-national interest in the destination, which is usually driven by a profit motive.

One of the significant issues related to the imposition of regulatory measures for ecotourism operators is the perceived loss of control in the delivery of services and decision-making. In a discussion of the adventure and ecotourism industry in South Carolina, Tibbetts (1995–6) illustrates that many of the operators in this region favour self-regulation and voluntary guidelines. There is a strong belief by such operators that since they are the ones out there working in the environment they know how to manage their affairs without government help, how to solve their own problems, and how to take the lead in terms of planning and implementing an appropriate product. This may very well be true, but there are other, more broadly based concerns that may need to be considered. The operation of an ecotourism business involves an interaction with other tourism businesses, other land-users, local people, tourists, and so on. The result is that the potential conflicts that may arise from these interactions (e.g. the overuse of trail systems, and the dumping of wastes in the Galápagos Islands) are not simply local and operator-specific, but rather have larger regional and socio-ecological implications. A key question to address is whether or not policy-makers can afford to provide the time for operators to distil-out problems of complementarity and conflict, especially when it is the resource base that should be protected first over the needs of tourists and private industry. We must consider that today, more than ever, businesses must be accountable to themselves, but, more importantly, to society. Given this axiom, it is appropriate that we continue to press the issue of whether or not the tourism industry and environment might better be served by regulation or self-regulation, or indeed both. In addition, research needs to concentrate on the difference and/or similarities between ecotourism and more conventional forms of tourism with respect to regulation and self-regulation, in an attempt to discern whether ecotourism businesses are in fact more ecologically minded than mass tourism firms.

## Agents and operators

Travel agents and tour operators are key stakeholders in the relationship that exists between the destination and the tourist. Both have a tremendous influence on tourists relative to the choices made and types of experiences gained from travel. To destinations eager to capture part of the tourism market, agents, although not the only means by which travellers learn about various destinations, must be familiar with the broad spectrum of opportunities available to the travelling public. Milne and Grekin (1992) and Wilkinson (1991) report that at least for some peripheral destinations, or those just starting to develop a tourism product, travel agents can play a critical role in counselling destination-naive tourists. The information profile of a destination – how firmly entrenched it is as a travel option – especially to the travel agent, therefore, places each destination in a position of competitive advantage or disadvantage.

Tourism opportunities are typically offered by operators, whom Metelka (1990: 110) defined as owners or managers of a place of business. Recently, however, service provision in ecotourism has included the 'outfitter', who has strong ties to outdoor recreation, where such providers have been providing adventure (e.g. hunting and fishing) experiences for tourists for some time (at least in North America). The types of experiences of interest to natural resource-based tourists (e.g. adventure tourists and ecotourists) have necessitated the inclusion of businesses that are prepared to offer equipment and other specialised services for tourists. Outfitters, then, are those 'commercial businesses that provide a person with the equipment necessary for an activity or experience' (ibid.: 111). As Tims (1996) observes, outfitting traces its roots back to the early explorers who provided a service to people, usually in a manner that placed humans against nature. More recently, outfitting has progressed to the point where it has taken on professional status through America Outdoors and the Professional Guide Institute (PGI). The PGI provides training for people in the

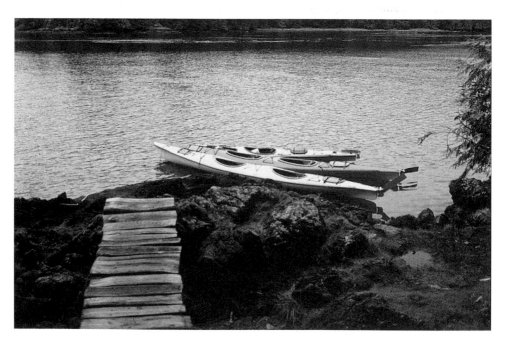

**Plate 6.1** The kayak continues to be a mainstay of many adventure-related operations in North America and is now used by ecotour operators to take tourists to many different settings

areas of wildlands heritage, back-country leadership, interpretation and outfitter operations; with the mission to 'identify, enhance and disseminate the natural interpretive and educational resource of the outfitting industry so that outfitters and guides can offer the highest quality of experience to the public' (ibid.: 177). Clearly, the PGI operates with a mandate that involves guides as a natural ingredient to providing effective back-country experiences.

On a more practical note, an excellent presentation of a hands-on approach to ecotourism operation and outfitting is provided by Mitchell (1992). Although this is too detailed to present in its entirety here, he identifies the following main programme-related aspects in structuring and running ecotours: professional tour guiding, a survey of personal assets, a survey of recreational assets, balancing demand, environmental education, tour planning, pricing, preparing the client for the trip, questions clients ask, packing lists, client medical information, cooking and food, conflict resolution, camp and camp cleanliness, first aid, legal considerations, on-the-trail sitings, and evaluation. For those more interested in actually organising and conducting tours, this document is a must. In addition, more specific guides related to first aid, risk management, leadership, interpretation, recreational programming, marketing and finance, and so on, are recommended to prepare the operator for the task of leading successful ventures in the field.

## Related research on operators

One of the earliest papers to emerge on the operation of nature tour companies was written by Ingram and Durst (1989), who designed a study to document empirically the promotion and activities of such companies in the United States. These authors used the Specialty Travel Index to identify 78 tour operators, of which 32 agreed to take part in the study. They concluded that firm size ranged from 20 to 3,000 clients, with three serving over 1,000 clients each. The majority of their client base was described as 'outdoor enthusiasts, retired couples or students, equally male and female, over 30 years old, and usually travelling as individuals in late summer' (ibid.: 12). The companies had been in operation for seven years, on average.

In other research, Eagles and Wind (1994) employed a content analysis methodology in analysing the advertising of 347 guide-led ecotours in 50 different countries, most of which were Canadian based. They found that on average the tours had a guide to participant ratio of about 1:13. This statistic seems to be consistent with the outdoor recreation literature, which recommends guide to participant ratios of approximately 1:12 for hiking (Ford and Blanchard 1993). It should be noted that in most cases ecotourism is dependent upon parks and protected areas and those areas' specific guidelines and policies. As such, many parks restrict party sizes to manageable numbers, which means approximately 10 or 12 individuals to a party. The reason for this is not only ecologically based (too much impact on certain areas), but also sociological, in that large groups often contribute to crowding, noise and other disturbances. Ecotour operators must be sensitive to such policies, and act accordingly. These potential impacts are well documented in the work of Hendee et al. (1990) and Hammitt and Cole (1987). For example, the latter authors suggest that 'large parties are typically defined in wilderness areas as groups larger than 8 to 10 members' (ibid.: 173). Hendee et al. (1990: 369) suggest that in most regions, parties with more than ten individuals account for only about 5 per cent of all groups, and illustrate that party sizes are declining owing to park restrictions and the 'larger, organized groups have generally become smaller as both managers and organization leaders have become concerned about the impact of large groups on the environment and on other visitors'.

In a survey of 24 North American ecotour operators, Yee (1992) developed a profile of the content and philosophical orientation of such operations. While most (63 per cent) had been in business for 2–15 years, 17 per cent had been operating between 15 and 20 years, many suggesting that they were offering ecotours long before the term originated. Ninety-two per cent of respondents followed a code of ethics, and 38 per cent considered ethical conduct in pre-trip orientation; 29 per cent addressed ethics in printed information packs; 13 per cent of those surveyed used videos or slides on ethics, while 33 per cent used lecture formats (see Chapter 9 for a more comprehensive look at ecotourism and ethics). The use of trained interpreters and naturalists was also addressed in Yee's research, where it was found that 75 per cent of the respondents had 'ecologists, naturalists or other experts on staff to aid in conducting their tours' (ibid.: 11).

Researchers have also analysed the degree to which ecotour operators promote and perceive themselves as being environmentally friendly (Weiler 1993). In this exploratory study, it was found that about 40 per cent of operators were promoted as being environmentally friendly; 66 per cent of operators felt that their tours were either beneficial or very beneficial for the environment (only one was said to be harmful or very harmful); 70 per cent said that their tour educated tourists about impacts; and only 7 per cent said that their tour enhanced the environment through, for example, removing others' rubbish.

Owing to the nature of the ecotourism experience, tour operators and resource managers have been forced into much closer proximity to each other, which, according to the research of Moore and Carter (1993), has not been a compatible relationship. In an interview of 16 nature-tour operators in Australia, the authors found that from the operators' perspectives, resource managers did not understand aspects of their business, including the profit motive, costs associated with marketing, costs of operations, the motivations and needs of tourists, and the need for minimal infrastructure in remote areas. Conversely, the operators had concerns related to the managers, including the fact that managers did not have a conservation ethic, failed to provide visitors with appropriate information, had little understanding of the need to control visitors, and did not understand resource fragility. These are fundamental differences that exist between the philosophies of the two camps. In particular, the profit motive is one that has not historically been of concern to parks, given park managers' concern for preservation over use. Although not considered in the paper by Moore and Carter, research needs to examine the relationship that exists between the ecotour participant (i.e. the individuals associated with a tour group) and the non-tour participant, especially from the perspective of the park manager. It would be interesting to understand how park managers view tour participants relative to the smaller group or individualised travellers. These authors suggest that a close working relationship must be developed between these two parties in order to meet the objectives of each group.

In a more comprehensive analysis, Higgins (1996) provides an overview of the structure of the nature tourism industry in the United States. In particular, he states that ecotourism is based on the existence of (1) outbound nature tour operators, located in large key cities in industrialised countries; (2) inbound nature tour operators, who are centred in non-industrialised countries and usually provide services in one country; and (3) local nature tour businesses, which include hotels, restaurants, ecolodges, souvenirs, guides, and so on. It is the outbound nature tour operators that play the greatest role, according to Higgins, in linking clients with other businesses and destinations around the world, particularly those in non-industrialised countries. In a survey of 82 US nature tour operators, Higgins found that such operations have increased in number by 820 per cent between 1970 and 1994 (from seven to 83). Higgins also discovered that client lists for nature tour operators ranged from 25 to 15,000, with an average of 1,674 – 35 of which served over 1,000 clients in 1986 – with the largest five serving 40 per cent of the total market.

# Professionalism

An aspect of ecotourism service provision that appears to be discussed more frequently in the literature is professionalism (see Wearing 1995, for example). Such discussions have emerged in other, related fields – including the recreation and leisure field, where debate has continued in the United States, according to Sessoms (1991), since the 1930s. (Currently, two organisations that drive certification efforts in the United States are the National Recreation and Park Association and the National Council for Therapeutic Recreation Certification.)

Professionalism is viewed in essentially two ways. In one sense, the term implies only that one is paid for a service (for example, a professional football player). It is, however, the stricter sense of 'profession' regarding conceptions surrounding occupations, such as lawyers and doctors, that is more difficult to define, and the focus of this section.

Foremost in the minds of professional groups and the lay public, the word 'professional' seems to be represented by two main themes, as suggested by Wilensky (1964: 138): '(1) the job of the professional is technical – based on systematic knowledge or doctrine acquired only through long prescribed training; and (2) the professional man adheres to a set of *professional norms*'. In recreation, researchers and practitioners, historically, found it difficult to agree on a set of professional criteria to guide accepted practice. Professions therefore need to satisfy, and be consistent with, broad criteria in bonding the discipline together, as evident in the following work by Weschler (1962):

1   The profession has a body of specialised knowledge.
2   The profession sets its own standards.
3   The profession requires extensive preparation.
4   Specialised knowledge is communicable.
5   Service is placed over personal gain.
6   The profession has a strong professional organisation.

In recent years, university and college programmes in Canada have shown interest in addressing the issue of professionalism in the adventure tourism and ecotourism fields. The adventure Travel Guide Diploma programme of the University College of the Cariboo (UCC) in Kamloops, British Columbia, is one such programme that has been developed to prepare a more highly trained labour force in this area (Olesen and Schettini 1994). These authors rely on projections that suggest that the demand for highly trained personnel in the adventure tourism industry will grow at 3.6 per cent (compounded rate), which, comparatively, far exceeds the anticipated 1.6 per cent yearly growth rates of the entire British Columbian labour force. This programme recognises the integral relationship between industry and government (and government-sponsored institutions) in certifying highly trained instructors, guides, and operators. The UCC programme emphasises the following five main themes as being critical to the success of the programme: (1) technical skills (e.g. wilderness first aid training), (2) business skills, (3) hospitality and service skills, (4) environmental skills, and (5) intimate involvement in industry associations (e.g. student networks with potential employers).

The tourism field will have to come to grips with the notion of professionalism (whichever form this takes), given the tremendous interest in this area of late. One of the key elements to overcome is enabling the tourism worker to feel personally competent and confident in his or her work; as Wright (1987: 17) points out, 'it is preferable to establish credibility through capability, competency, efficiency and a professional attitude. One earns respect, one does not establish respect through a glossy diploma on the wall.' Accreditation, certification and guide traininng are the main aspects by which to demonstrate professionalism in ecotourism.

## Accreditation and certification

Along with regulation (as outlined on pp. 114–15), there has been an increased focus on the relevance of certification and accreditation in the tourism industry over the past few years. Certification and accreditation programmes act as the basis for the education and the provision of skills of those working in the industry. The heightened level of attention afforded to certification and accreditation is discussed by Morrison *et al.* (1992), who identify a number of existing programmes in North America, including programmes for the posts of Certified Travel Counsellor, Certified Hotel Administrator (the two longest-standing programmes), Certified Tour Professional, Certified Meeting Professional, Certified Hotel Sales Executive, Certified Festival Executive, Certified Incentive Travel Executive, Certified Travel Marketing Executive and Certified Exhibit Manager. According to Morrison *et al.* (1992), the two terms 'certification' and 'accreditation' may be differentiated on the following basis:

- *Certification.* A process by which an individual is tested and evaluated in order to determine his or her mastery of a specific body of knowledge, or some portion of a body of knowledge.
- *Accreditation.* A process by which an association or agency evaluates and recognises a programme of study or an institution as meeting certain predetermined standards or qualifications. It applies only to institutions and their programmes of study or their services.

As Morrison *et al.* (1992) point out, certification deals with the individual professional, while accreditation is concerned with programmes and institutions. An interesting conclusion of their research is that certification programmes have grown to replace the university as a means by which to gain expertise and professionalism in the tourism industry. They caution, however, that although certification will indeed increase in the future, it should not be a substitute for travel and tourism degrees. In addition, whereas universities are able to offer a good grounding in the theoretical aspects of tourism, they may also diversify into offering certification options to their graduates through their programme offerings. Outside tourism, the relevance of accreditation has been debated in several other disciplines. Scalet and Adelman (1995), for example, discuss the pros and cons of accreditation in fisheries and wildlife programmes, while Silverman (1992) examines accreditation of undergraduate programmes in environmental health science.

In the realm of adventure recreation, there has been significant debate on how best to certify certain aspects of the adventure programme experience, including standards of programme quality, professional behaviour, professional conduct, and risk management (Gass and Williamson 1995). Similarly, these have revolved around the certification of leaders and guides through organisations such as the American Mountain Guides Association and the Wilderness Education Association, while the American Camping Association and the National Recreation and Parks Association have implemented accreditation programmes to meet their own specific needs. Two studies found overwhelming support for accreditation over individual certification as it (accreditation) provided an umbrella to ensure a higher degree of professional credibility and less reliance on individual operators (see Cockrell and Detzel 1985; Bassin *et al.* 1992). These findings were considered in the recent development of a US national accreditation process that focuses on programming in the adventure recreation field by the AEE (Association for Experiential Education 1993), which is designed to consider the technical (hard) skills of adventure recreation, but also the teaching and ethical elements that need to be consistent from the programme side of the experience. This document provides a means by which to systematically evaluate operations in areas related to ethics, risk management, staff qualifications, transportation and technical skills. For example, with respect to the environmental aspects of river rafting, code 43.B.01 suggests that 'Staff are familiar with the operating sites for rafting. *Explanation*: A pre-site investigation is conducted to understand the conditions as well as the education/therapeutic possibilities. This includes locating appropriate takeout areas and access to safe transportation sites' (AEE 1993: 82).

According to Gass and Williamson (1995: 25), this national programme maintains the following accreditation procedure:

1  *Self-assessment and documentation stage*. Operators must compare their programme against the AEE programme guidelines.
2  *Verification stage*. A team of between two and four reviewers visit the operator and conduct interviews, inspect equipment, and observe activities as necessary. A standards report card is used to determine whether a component of the programme 'passes', 'fails', is 'conditional' based on the need to remedy the situation, or does not apply in the case where an activity is not associated with a programme.
3  *Follow-up stage*. Findings from the verification stage are forwarded to the Program Accreditation Services committee, who send a recommendation to the AEE Executive Board, who in turn decide whether the programme should receive accreditation.

In further support for the idea of accreditation, Gass and Williamson (1995: 23) offer the following:

1  Accreditation provides adventure programmes with the ability to achieve standards without losing the flexibility to decide and design how these standards are met.
2  Accreditation takes a systematic view of the process of adventure programming rather than dividing it into individualised categories.
3  Accreditation encourages ongoing improvement through internal and external review.
4  Accreditation assures clients, agencies, and resource managers that a programme has clearly defined and appropriate objectives and maintains conditions under which their achievement can be met reasonably.

As suggested in Chapter 2, ecotourism shares some aspects of adventure recreation/tourism, more from the context of setting than in programming. This element of setting (e.g. tropical rainforests) may provide a component of risk that is absent from other, more conventional forms of tourism. To a lesser extent than adventure travel, ecotourism may be seen as bridging a gap between various forms of adventure recreation and tourism. In comparison to outdoor recreation, however – in particular, outdoor guides and leaders – tourism lags far behind, and as in adventure recreation, programming in ecotourism will be a key issue in the future.

Recently the topic of accreditation and certification has been debated in ecotourism, consistent with this trend in the adventure pursuits industry. For example, Wearing (1995: 34) suggests that professionalisation and accreditation in ecotourism will continue to be at the forefront of discussions related to regulation and control, as a means by which to provide focus to an industry that is almost impossible to limit (in terms of expansion). He feels that accreditation affords the opportunity to improve tourism industry standards while at the same time ensuring high-quality services and programmes in a very competitive marketplace. Wearing identifies a series of advantages and disadvantages of accreditation as they relate to ecotourism, as follows:

## Advantages of accreditation

1 The ecotourist has the knowledge of what will be taught, and the approaches used will be the best and the safest.
2 The ecotour operator has a recognised and accepted methodology and thus increased social status.
3 The employer of ecotourism guides has knowledge of an accepted industry standard.
4 The government has the knowledge that new operators have been taught to a minimum standard

## Disadvantages of accreditation

1 The idea that one needs a service which is best provided by an expert creates a relationship of mutual dependency and a social distance between the ecotourist, the operator, the host community, and the natural environment.
2 Ecotour operators are assessed on their accomplishments rather than their human characteristics to the effect of depersonalising the human–natural environment relationship.
3 Professionals can tend to rationalise and focus on facts, objective data, and procedures, thus potentially losing the intrinsic, intuitive association so often formed through ecotourism.
4 Through commodification, the operator may come to view the natural and cultural environments as a means to an end and so become alienated from the natural world and hence unable to encourage the ecotourist or encourage environmental awareness.
5 Loss of perceived freedom by the individual ecotourist, who feels he or she must do as the operator says; this restricts the ecotourism experience.
6 The imposed structure that is required to assess or measure the ecotour operator entails limitations.

The message put forward by Wearing is that although accreditation would certainly reduce risk, increase standards, and increase status, there is a danger of restricting innovation and accessibility within the ecotourism industry. One of the significant dilemmas, therefore, is the degree to which the industry can afford to emphasise innovation and accessibility at the expense of ensuring that proper standards are being upheld. The issue is one that clearly requires more time and education for operators in the field. In addition, it is a normative concern as the likelihood for compliance is directly related to how operators feel individually, but also collectively, about the process of accreditation. If it is imposed upon them, only time will tell how operators will react to the imposition. However, in time accreditation may be fully accepted as proper or appropriate business practice.

A good example of the accreditation process in action can be found in the Australian National Ecotourism Accreditation Program (now referred to as the Nature and Ecotourism Accreditation Program), where those working in the field (i.e. operators) submit to a standardised procedure. At present, operators are given a core level of accreditation if they satisfy the programme's basic criteria; however, the system encourages operators to implement measures beyond the standards of the core criteria in earning advanced standing. Eligibility for accreditation is based on eight criteria (Chester 1997: 9). According to these criteria, ecotourism:

1 focuses on personally experiencing natural areas in ways that lead to greater understanding and appreciation;
2 integrates into each experience opportunities to understand natural areas;

**Table 6.5** *Ecotourism accreditation fees in Aus$ (from November 14, 1996)*

| Fees – based on annual business turnover ($) | 0–100,000 turnover p.a. | 100,000– 250,000 | 250,000– 1,000,000 | 1,000,000– 3,000,000 | 3,000,000– and over |
|---|---|---|---|---|---|
| Ecotourism accreditation document fee (per. doc): | | | | | |
| Members | 75 | 75 | 75 | 75 | 75 |
| Non-members | 175 | 175 | 175 | 175 | 175 |
| Application fee (only in year 1) | 75 | 100 | 175 | 250 | 400 |
| Annual renewal fee | 100 | 150 | 300 | 500 | 750 |
| Total fee in year 1: | | | | | |
| Members | 250 | 325 | 550 | 825 | 1225 |
| Non-members | 350 | 425 | 650 | 925 | 1325 |

*Source*: McArthur (1997)

3 represents best practice for ecologically sustainable tourism;
4 contributes to the conservation of natural areas;
5 provides constructive, ongoing contributions to local communities;
6 interprets, involves, and is sensitive to different cultures (particularly indigenous ones);
7 consistently meets client expectations; and
8 its marketing is accurate and leads to realistic expectations.

In Australia there was resounding support for the idea of ecotourism accreditation in the mid-1990s (McArthur 1997) which culminated in the launching of the NEAP at Australia's 1996 national meeting. Accreditation is offered to those operators working in the accommodation, tour operations and attractions sectors, and operators are accredited by first completing an application (through a nomination process involving three referees), and by paying a one-off application fee and an annual fee. Table 6.5 provides an overview of this fee structure.

One of the key features of the Australian accreditation initiative is the level and style of enforcement employed in the programme. Each of the applicants is monitored in four ways: (1) the applicant's honesty in completing application documents; (2) feedback from clients; (3) feedback from referees (random selection of operators); and (4) through random audits.

The most recent change made to NEAP is the inclusion of a nature tourism category allowing the vast number of nature tour operators in Australia to receive accreditation. These nature-based and sustainable-oriented operators, who do not fit the stricter definition of ecotourism as outlined in Australia, may thus become more responsible operators according to the guidelines of the NEAP initiative. The other positive change to NEAP is the recognition that programmes ought to be accredited rather than operators. This means that although operators may run a number of programmes, it is only perhaps one or two, for example, of these programmes which are worthy of accreditation status. This gets away from a more broadly based operator-specific method which was not necessarily representative of the ideals of ecotourism within all programmes.

A more recent view of certification and accreditation is the notion that both concepts are not equal entities, but rather that accreditation is a more all-encompassing or higher status label which acts as an umbrella for those who may be certified. As suggested by Honey and

Rome (2001), a certification programme certifies the certifiers. These authors identify two main types of certification methodologies as follows: (1) process-based environmental management systems, which are driven by environmental management systems such as environmental auditing and environmental impacts assessments; and (2) performance-based methods which use environmental, socio-cultural, and economic criteria, standards or benchmarks. As many of these concepts are relatively new to tourism, it is worth while to provide some basic definitions of terms which are little understood (see also Font and Buckley 2001; Issaverdis 2001). This terminology is further elaborated as follows (all from Honey and Rome 2001):

- *Environmental Management System.* The EMS is part of the overall management system that includes the organisational structure, responsibilities, practices, procedures, processes, and resources for determining and implementing the environmental policy. An environmental management system includes tools such as environmental impact assessment, environmental auditing and strategic environmental assessment.
- *Audit.* Is a systematic, documented, periodic, and objective evaluation and verification of how well a particular entity (company, product, programme, individual, destination, etc.) is doing compared to a set of standards.
- *Benchmarking.* Is the process of comparing performance and processes within an industry to assess relative position against either a set industry standard or against those who are 'best in class'.
- *Best practice.* Is used to designate highest quality, excellence, or superior practices in a particular field by a tourism operator. It is widely used in many award and certification programmes, as well as in academic studies, to designate best in a particular class or a leader in the field. 'Best', however, is a contextual term. There is no set standard of measurement and the term is often loosely or ill-defined. (As suggested by Issaverdis, 2001, there is little difference between best practice and benchmarking.)
- *Ecolabelling.* Describes a scheme in which a product or service may be awarded an ecological label on the basis of its 'acceptable' level of environmental impact. The acceptable level of environmental impact may be determined by consideration of a single environmental hurdle or after undertaking an assessment of its overall impacts. (See also Font and Buckley 2001.)
- *Ecotourism certification.* Programmes that cover businesses, services and products that describe themselves as involved in ecotourism. They focus on individual or site-specific businesses, have standards that are tailored to local conditions, and are largely or totally performance-based. (Buckley, in press, notes that ecocertification in tourism is a form of ecolabel in the international trade sense.)

The importance of these programmes to ecotourism was noticeable in the recent (May, 2002) World Ecotourism Summit in Quebec city, Canada, where a number of researchers and organisations gathered to discuss the fate of such programmes. As Buckley (in press) notes, accreditation and certification has become quite politicised because of the sheer number of such schemes in operation (upwards of a hundred), and the associated control over those which are most prominent. Buckley suggests that the World Ecotourism Summit was used as a platform by Green Globe 21 and the Ecotourism Association of Australia to launch a draft International Ecotourism Standard for certification, which he observes is the Australian NEAP process offered in other countries through Green Globe 21. Where this leaves other schemes is an immediate point of interest, along with the associated implications such a monolithic unit might create. Just as important in the debate over the significance of such schemes is the basic question of whether accreditation schemes are worth all of the energy put into them. Buckley notes that they may be less effective than simple government regulations such as codes and laws within parks. At present, it remains that these schemes are politically charged and have an uncertain future, and the debate over their utility, as well as nomenclature, promises to be an interesting one.

## Guiding

The literature on guiding in the outdoor recreation and adventure tourism is, comparatively, more abundant than what has emerged in the ecotourism field to date. Despite the relative dearth of information, there is the understanding that in practice there is little that is more important to the safe and rewarding participation in an ecotour: the value of a good guide to the overall success of an ecotourism venture cannot be underestimated. Defined, an ecotour guide is 'someone employed on a paid or voluntary basis who conducts paying or non-paying tourists around an area or site of natural and/or cultural importance utilizing ecotourism and interpretation principles' (Black *et al.* 2000: 3). These are individuals who must possess a great many skills in the effective development and presentation of the trip. Priest (1990) suggests that guides need a number of broad skills which can be loosely organised into hard, soft, and meta skills. The first of these include the solid and tangible skills that make guides effective. Some of these might include technical skills (e.g., paddling or climbing competencies), safety skills (e.g., advanced wilderness first-aid) and environmental (e.g., minimum impact camping skills). Soft skills are those which are organisational (e.g., planning, preparing and executing skills), instructional (e.g., appropriately teaching how to paddle) and facilitation (e.g., those skills which foster productive group dynamics). Meta skills are explained by Priest as those related to decision-making abilities, problem-solving, experience-based judgement, and the knowledge of when to employ various styles of leadership, including democratic (allowing the group to make certain decisions), autocratic (taking the lead in emergency situations), and abdicratic (abdicating or delegating responsibility to the group). In view of the fact that ecotourists are a discerning group who may be quite vocal and active in environmental matters, guides and other front-line employees must be educated on natural history as well as any other issues related to environmental policies of the region and methods which have been or are being employed to ensure preservation of species or landscapes (see Mitchell 1992). A study by Weiler and Davis (1993) on the roles of nature-based tour leaders reported that the main roles of the tour leader (on a five-point scale; five being most important), were organiser (mean = 4.6), group leader (4.5), environmental interpreter (4.3), motivator (4.2), teacher (4.0), and entertainer (3.4). As Weiler and Davis suggest, in reference to the leader:

> S/he must be an organizer, a group leader, a teacher, and even an entertainer. In nature-based tourism, the tour leader must also be responsible for maintaining environmental quality, by motivating visitors to behave in an environmentally responsible way during the tour, and by interpreting the environment in such a way as to promote long-term attitude and behavioural change.
>
> (Weiler and Davis 1993: 97)

More recently, Weiler and Ham (2001) note that in Australia, which is perhaps the most advanced ecotourism country in the world, there still does not exist industry-wide legal specifications regarding the rights, working conditions, pay, and so on, of ecotour guides. Although the recently developed Ecoguide Certification Program (a voluntary, industry-led certification programme developed by the Ecotourism Association of Australia), does not fully address these issues, it has come about with the aim of providing a qualification that is designed to reward guides who achieve various competencies. The programme is not designed to train guides, but rather to assess their skills, knowledge, actions and attitudes

**Table 6.6** *EcoGuide programme benefits*

*Benefits to nature and ecotour guides*

1  A recognised industry qualification.
2  An opportunity to promote guiding services as genuine nature/ecotourism.
3  A defined competitive edge rewarded through factors such as better job opportunities.
4  Access to relevant, appropriate and reduced-cost training materials/networking options.

*Benefits to nature and ecotourism operators*

5  A simple method of recognising and recruiting quality guides.
6  Identification of training gaps/needs within operations.
7  Improved guiding practices leading to fewer negative environmental and cultural impacts and increased client satisfaction.

*Benefits to nature and ecotourism consumers*

8  An assurance of guides that are committed to providing quality nature or ecotourism experiences in a safe, culturally sensitive and environmentally sustainable manner.

*Benefits to protected area managers*

9  Improved guiding practices that lead to fewer negative environmental impacts.
10  Guides who role model and ensure good environmental/cultural behaviour.
11  A framework of standards applicable to interpretive rangers.

*Benefits to the environment*

12  Guides providing relevant and appropriate interpretation that inspires clients and workplace employees and encourages minimal impact actions and a conservation ethic.

*Source*: Crabtree and Black (n.d.)

(Crabtree and Black n.d.), with an overall purpose of instilling a philosophy of best practice among practitioners. Table 6.6 provides a comprehensive overview of some of the benefits which may accrue to a number of different tourism stakeholders, as outlined in the EcoGuide Program (pp. 8–9).

Clearly the dearth of information on ecotour guiding must be addressed through programmes like the one mentioned above, and through more intensified research. The implications of a better organised and trained guiding workforce are global, as ecotourism continues to be a force for local economic development just about everywhere. One of the sad realities of ecotourism, despite its overarching philosophies, is that it at times concentrates economic impacts into the hands of a few, especially in the lesser developed countries. Guiding is one of the aspects of ecotourism that contributes directly into the hands of local people, who may have little stake in the overall operation of the organisation. This has prompted Ham and Weiler (2000) to outline the following six principles for tour guide training which are designed to help countries of the lesser developed world develop sustainably through the ecotourism industry:

1  The initiative for training should come from the host country and ownership should remain with the host country.
2  Training content and methods should be informed by the literature on what constitutes good or best practice ecotour guiding and the adult training literature, with appropriate customisation to meet local needs and the target trainee group.

3   Training efforts must be systematically evaluated, and lessons learned from these evaluations must be documented and disseminated widely.
4   Training can and should be made accessible, both logistically (location, time and cost) and intellectually to residents in peripheral and rural areas.
5   Delivering cost-effective guide training by mobilising resources for multiple 'ends' is an important sustainable development strategy.
6   Training and supporting in-country trainers is essential for ensuring that ecotourism is of benefit to host economies.

The specific implications of guiding in general may be examined in the following section on interpretation (as an example of an ecotourism competency), and more broadly in the following chapter on programme planning for ecotourism.

## Interpretation

While interpretation is often thought of solely in the context of persons working within parks and protected areas (interpreting the landscape), according to Barrow (1994) interpretation is far more elaborate, and based on three types of planning: (1) town and country, (2) marketing, and (3) education. In the town and country approach, appropriate land use philosophies and techniques are used to protect valuable land from development, provide access to the public, and to help create an effective tie with the resource base. Marketing interpretation, on the other hand, deals with how best to understand various user groups and their particular wants and needs. The marketing approach to interpretation is often product oriented, but must temper this with other concerns of the site (e.g. resource protection). Finally, Barrow suggests that interpretation employs educational theory in its plan development in order to understand how people learn, and what to teach them.

In order to manage effectively, interpretive planners must be sensitive to the learning, behavioural and emotional sides of interpretation. Veverka uses the following example of an archaeological site to illustrate the relationship between these three objectives in interpretation:

> The Learning Objectives are designed to provide basic topic information or understanding such as: 'The majority of the visitors will be able to describe three reasons why protecting archaeological sites benefits all visitors'. But the real – or most important – aim the manager may have in mind for interpretation to accomplish is to prevent visitors from picking up 'souvenirs' such as pottery shards at archaeological sites. So the Behavioural Objective might be: 'All visitors will leave alone any artifacts that they may find at the site and not to take them home'. It is the job of the Emotional Objectives for the interpreter to get the visitors to appreciate the value of artifacts left in place, and feel that they are doing a good thing by not removing anything. So an objective of this kind might be: 'The majority of visitors will feel a sense of responsibility for not touching any artifacts they may find on the ground'.
>
> (Veverka 1994: 18)

The behaviour is the action that the interpreter would like to occur as a result of the interpretive programme, whereas the emotional objective prompts the interpreter to strive to realise how best to get the visitor to feel as though he or she is behaving in the appropriate manner at the site and beyond; that is, that the behaviour is something that they should want to do. Veverka illustrates that the same philosophy is used by advertisers today, in that they make you feel as though you want or need a particular product. The behavioural response is to actually go out and purchase a product.

**Plates 6.2 and 6.3** Experienced multilingual interpreters are often central to the ecotourism experience

Interpretation, as identified earlier by Weiler and Davis (1993), plays a significant role in the delivery of ecotourism services. For example, on Great Barrier Reef Marine Park in Australia, Hockings (1994) examined the tour operator's role in marine park interpretation. While staff–recreationist interactions are not very high as a consequence of the level of use and size of the park, tourists have come to rely on the various tour operators in securing information. Of a sample of 170 operators, 72 per cent of respondents indicated offering interpretation as part of their programme. A higher proportion of this number were found to be day reef trip, diving and sailing operators, while a lower percentage of fishing operators provided interpretation. A significant finding of the study was that 56 per cent of staff involved in interpretation had no relevant formal qualifications, while only 4 per cent of all staff had formal qualifications. It was suggested that this overall lack of training could not be linked to type of activity offered (types of activities included scenic flights, fishing, camping, day reef trip, seaplane rides, dive trips, and so on), as those indicated as having the appropriate interpretive qualifications were evenly spread across operator types and sizes. Operators, according to Hockings, relied more on practical experience as a means by which to inform clients.

Forestell (1993) has developed a unique model of interpretation and environmental education that he operationalised in his work in whale-watching. This author suggests that tourists experience different cognitive states during a whale-watching excursion that require the naturalist/interpreter to articulate information in three distinct periods: pre-contact with whales, contact, and post-contact. Pre-contact information involves learning on how participants should interact with the whales. This may be supplemented with geographical information and marine environmental education. In the second, or contact, stage, interpreters focus on the need for participants to seek answers to a variety of questions related to the observation of the animals. Questions typically relate to whale identification, whale behaviour, verification of knowledge, and safety. The goal of this stage is to 'generate motivation to learn by creating or uncovering an imbalance between an individual's initial knowledge base, and some current perception of the world . . . [and] provide sound knowledge to allow the participant to regain cognitive balance' (Forestell 1993: 274). The final stage, post-contact, is characterised by the participants' need to make pre- and post-trip comparisons of whales and the marine world, and in addition to consider broader environmental issues (including active and financial support for conservation programmes or issues). Clients also question leaders about the well-being of the whales. Forestell suggests that ecotourism experience has the ability to empower people to synthesise unscientific observation with scientific fact, which he feels is not an inferior substitute for hard data, but rather one that is experiential and practical. This model demonstrates the need for interpreters to be sensitive to the different stages of an ecotourism experience. In situations where guides are providing more personal knowledge of a setting only, Forestell perhaps would argue for more of a trained approach to the delivery of the interpretive product, which would ultimately enable the tourist to benefit further from the experience.

## Conclusion

This chapter has illustrated that sound planning and management related to policy, regulation, and accreditation, certification and professionalism are necessary considerations for the future viability of the ecotourism industry. While some governmental bodies have recognised early the importance of government leadership in the ecotourism industry, others have not been so quick to act. There must be both leadership and partnership between government and industry in the development of mechanisms to control the impact of ecotourism on the environment, and to ensure that operators are providing safe and enriching programmes to their clientele. Such mechanisms will most likely include policy, regulation, accreditation and professionalism in their various forms.

## Summary questions

1 What is ecotourism policy, and how might it create a better ecotourism industry?
2 How do accreditation and certification differ, and how might they contribute to a heightened level of professionalism in the field?
3 One of the key defining characteristics of ecotourism is learning about the environment. How is it that ecotourism guides are able to change behaviours through emotional objectives?
4 How has the Australian accreditation programme been changed most recently to take into consideration nature tourism? What might the positive and negative implications of this move be to the ecotourism industry that country?
5 Explain why a good combination of soft, hard, and meta skills are needed by ecotourism guides.

# 7 Ecotourism programme planning: a focus on experience

One of the mysteries in tourism research, at least to the author, is the absence of a more thorough treatment of the recreation and leisure studies in the literature. By this it is suggested that although recreation and tourism can be viewed as separate fields of study, there is quite a strong conceptual connection between the two areas. This has been emphasised by a number of authors over the years. For example, Jansen-Verbeke and Dietvorst (1987) and Edginton *et al.* (1980) suggest that in analysing the leisure discipline from the perspective of the individual there is little difference between elements of leisure, recreation and tourism. Tourism therefore is very much a recreational beast, leading some to suggest that the basic motivation for tourism is the human need for recreation (Graburn 1989). This link is further emphasised by Metelka (1981), who defines tourism as 'Free spontaneous activity; synonymous with recreation. An activity done for its own sake, rather than for economic gain' (p. 90). Subsequent to this Metelka (1990) defined tourism in three ways: (1) the relationship and phenomena associated with the journeys and temporary visits of people travelling primarily for leisure and recreation; (2) a subset of recreation; that form of recreation involving geographic mobility; and (3) the industries and activities that provide and market the services needed for pleasure travel (p. 154). (See Mieczkowski, 1981, for a discussion of the relationship between leisure, recreation and tourism.) The resultant overlap between recreation and tourism is a topic in the work of Hall and Page (1999), who say that there is increasing convergence between the two concepts in terms of theory, activity and impacts. Perhaps one of the key reasons for differentiation is the notion that tourism continues to be examined from marketing and business contexts, while the recreation delivery system still maintains a strong focus on satisfaction (i.e., how the recreation delivery system might better satisfy the needs of people). In this chapter, an attempt is made to bridge the recreation and tourism gap through a process referred to as programme planning. This broadly based procedure provides the means by which to organise a series of steps, strategically, in piecing together better and more holistic experiences for ecotourists. In this regard, the focus sits squarely on the tourist.

## Why are we in business?

Perhaps one of the most fundamental differences between recreation and tourism is what might be referred to as a product mentality. The product mentality carries with it, as either perception or reality, an emphasis on profit and production, and marketing and business planning. Products, as we well know in the tourism literature, may be viewed as goods or as services, where goods are produced and services are performed. Tourism providers often view their offerings as goods in the sense that they are able to stockpile them for use again and again, while tourists view the product more as a service in the way in which experiences

Visitor Satisfaction
in Recreation

Visitor Satisfaction
in Tourism

Participant

Client

**Figure 7.1** Satisfaction in recreation and tourism

are immediately consumed on site. The tourism-as-service perspective is one advocated by Holloway (1994), who notes that tourists cannot inspect their package tour before the purchase. It is therefore a speculative investment involving a good deal of trust on the part of the purchaser. As regards tourism, there appears to be an overriding emphasis on the profit motive, which is likely a natural by-product of the marketing and business culture that pervades tourism. (The literature supports a vast array of tourism development examples that discuss the value and benefit of tourism almost solely on the basis of economic impact. The same holds true for ecotourism, where there is an ever-expanding base of literature which questions the interests and values of those involved in the business side of ecotourism (see Munt and Higinio 1993; Steele 1993).) This is not only a focus that stems from the private/commercial sector, but also one which has infiltrated the public realm. One of the classic examples is the municipal perspective where decision-makers have become quite adept at promoting tourism within their cities to bolster their economies. There is certainly nothing wrong with this type of behaviour. Problems arise, however, when the profit motive becomes the sole overriding motivation of service providers, with little regard for participant satisfaction.

One of the foundational aspects of recreation is the notion that service providers are in the business of facilitating, as much as possible, participants (as something different than clients) to have satisfying recreational experiences. Indeed, this is the recreational programmer's prime mandate, over all else. This is articulated in Figure 7.1, which, although a generalisation, contrasts tourism and recreation along the lines of satisfaction. The figure shows that in the case of recreation, the satisfaction of the visitor is central. Conversely, the profit motive is more central in tourism, in general, and client satisfaction is less a concern.

The accepted methodology for ensuring a common approach to recreational satisfaction over the broad breadth of recreational experiences is recreation programme planning. Programming is a process which entails following many simple but well-planned principles, which are not only geared towards the satisfaction of participants but also to better understanding how the agency may be accountable to itself in addressing issues related to capital, equipment, effective use of time and the behaviour of employees or clients. Simply, but effectively stated, programming is the process of organising resources and opportunities for other people for the purpose of meeting their leisure needs (Searle and Brayley 1993: 214). These needs may be educational, cognitive, affective, social, or psychomotor, and the satisfaction of such enables these individuals to live healthier, happier lives. Recreation programming is also a form of planning, following a typical strategic and systematic focus: dynamic, cyclical, interactive between people and resources, and seeking to implement and evaluate in order to achieve goals and objectives. The importance of programming to recreation is illustrated by Russell (1982), who states that:

The program is what recreation services are all about. All else – personnel, supplies, areas and facilities, budgets, public relations – exists primarily to see that the program occurs and that people enjoy participating. Planning is the tool that makes programs happen.

(Russell 1982: ix)

In general, the inherent strength of programme planning, as suggested above, is that it is one integrated process, broken down into a few broad units, which typically include philosophy, planning, implementation and evaluation. Depending on the model chosen, these units will specify certain aspects of the programming process. However, because there are literally dozens of tasks which need to be accomplished along the way, certain tasks will need to be done in a logical order, concurrent to others, or even in reverse order (due to the nature of the programme). This means that although there is a fairly common process to follow, certain tasks are not necessarily exclusive to one broad unit over another. For example, in some models risk management is placed in the programme implementation unit, whereas in others it may be part of the planning process of the programme.

It has been suggested that ecotourism is a form of tourism which is ethical, responsible, small scale, and community-based. Consequently, this type of tourism may be useful as an instrument to underscore the human, service values – elements which are central to the development of sound recreational programming. As this book has attempted to point out, ecotourism is far from perfect, especially in terms of how it is practised. This means that there is a great deal more that we could do in how we present our programmes to the ecotourism public. Programming can help in this regard by the development of a more standardised process of programme delivery which underscores many of the participant-based procedures which are underscored in recreation. And while organisations intent on accreditation are madly attempting to capitalise on values from the macro perspective (see Chapter 6), programme planning allows for the development of sound, ethical programmes at the individual or micro level, whether tied into broader schemes or not.

This means that how we plan for ecotourism should be drawn from a focus hinged on a participant satisfaction end, as outlined in Figure 7.2. If we plan with a profit end in mind, as outlined in the first case in Figure 7.2, the things we need to accomplish in realising that end – those which will need to go into the boxes in both cases of the figure – will be very different. The figure, although basic, thus serves to provide a simplified notion that the processes that we need to operationalise are very much dedicated towards achieving desired ends, whatever these may be.

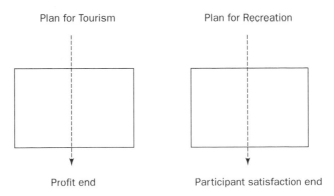

Figure 7.2 Planning for tourism and recreation

**Plates 7.1 and 7.2** No matter where the ecotour occurs, programmers must be careful to place the needs of the participants above all. In ecotourism these needs correspond to environmental awareness as well

Tourism employees coached in proper techniques of programming may be witness to the many tangible benefits which are generated for the programme and service provider, for other employees and, most of all, for the client. Also, programmers, who may include programme planners, guides, interpreters, and other service providers, and who may work in companies with as few as two people working in them, must often 'wear many hats'. This may mean involvement in administration, public relations, community liaison, marketing, promotion, evaluation, programme design, establishment of goals and objectives, needs assessment, leadership, budgeting, facilitation, teaching and environmental education, scheduling, and so on. Because the responsibilities of the programmer run deep, it is essential that, in keeping with the prime directive of participant satisfaction and the philosophy of the organisation, they must attempt to adhere to the following principles, adapted from Edginton *et al.* (1980):

- Placing the needs of the participant first.
- Commitment to the ideals of ecotourism.
- Protection of the participant's rights.
- Acquisition of knowledge prior to engaging in professional activities.
- Practice of the highest standards of professional service.
- Continuous upgrading of professional knowledge, skill and ability.
- Operating ethically and equitably.
- Maintaining a collaborative relationship with the participant.
- Operators must self-regulate their behaviour accordingly.
- Contributing to the development of the profession and other professionals.

These guidelines underscore the importance of a more ethical and humanistic approach to the delivery of services, where the customer is valued as more than simply a means to a financial end. Valuing and practising equity, high standards, collaboration with others, commitment, knowledge, and protection of rights is critical for success, but also in terms of the communication of such values within the operation.

# Programme planning

In fitting with the previous discussion, the following model is suggested for ecotourism programme planning (i.e., the movement from needs assessments to evaluation, in the context of a philosophical outlook) (see Figure 7.1). Table 7.1, although not exhaustive, provides an overview of many of the aspects of ecotourism programme planning, as they apply to each section. In the interests of space, only a few of these will be discussed in brief. (See Fennell, 2002b, for a more in-depth overview of this process.)

## Philosophy, mission, goals and objectives

The development of an ecotourism philosophy (for the industry, for a service provider, or for programming) demands the examination of some basic assumptions regarding the place of ecotourism in the context of the environment, the nature of human activity, the basis of human nature, and the nature of human relationships (Malloy and Fennell 1998a). These may further evolve into a series of other questions related to the meaning of leisure for individuals and for society as a whole. Kraus and Curtis (1990) write that the philosophy which leisure service organisations develop is a direct reflection of the acceptance of leisure as an important part of life and the extent to which it meets human needs. These needs may relate to accessibility, happiness from involvement in the pursuit, the consideration of a means–end

**Table 7.1** *Some important aspects of programme planning*

| Philosophy | Needs and assets | Planning | Implementation | Evaluation |
|---|---|---|---|---|
| Philosophy | Understanding | Programme structure | Green marketing | Why evaluate |
| Mission | needs | Brainstorming | Quality | Who evaluates |
| Vision | Defining need | Briefing | Staff training | Formative |
| Goals | Needs assessments | Health and safety | Public relations | evaluation |
| Objectives | Assets inventory | Transportation | Budgeting | Summative |
| Programme | SWOT analysis | Food and water | Itineraries | evaluation |
| theories | PESTE analysis | Clothing/equipment | Implementation | Evaluation models |
| Programme | Understanding the | Permits | Strategies | Programme |
| strategies | environment | Leadership | Operating in the | decisions |
| Programme | | Risk management | environment | Debriefing |
| approaches | | Interpretation | | Accreditation |
| | | Business planning | | Audits |

relationship, rights of the individual, freedom, as well as equity (see Edginton *et al.* 1998). An agency's philosophy, then, is a direct reflection of what the entity values. There are perhaps a series of other core environmental values which would be articulated by ecotourists over other types of tourists, as being fundamental to the travel experience. These may include reducing, reusing and recycling; harmony; exploration; multiculturalism; preservation of landscapes; biodiversity conservation; integrity; learning; service and knowledge.

Mission statements are important as vehicles by which to articulate the philosophical direction of an organisation. These should be bold and inspiring, and, to be effective, must communicate direction to customers and workers within the organisation. Campbell *et al.* (1990) suggest that mission statements should be built around a purpose, values, strategy, and standards and behaviour. In essence, all of the various things that the organisation will plan on doing will emanate from the mission statement. Just as people's philosophical positions are subject to change, an agency's mission statement may be too. Kraus (1997) writes that the mission should be reviewed periodically in light of changing conditions, needs and agency capabilities.

Kraus (1997) further writes that goals and objectives, although often used interchangeably, are different in what they aim to accomplish. Goals are designed to reflect the mission of an organisation, and may be viewed as purpose(s) which are often communicated as statements which emanate from the mission statement. On the other hand, objectives are much more specific, measurable, and attainable, depending on the nature of the service provider. They are the steps required to reach specific goals, and are written in a straightforward way, according to the actions to be achieved as well as the outcomes which are expected over a given period of time. Objectives will quite naturally flow from the various goal or purpose statements which are a direct reflection of the agency's overall mission.

Objectives are of two types. The first include programme (also called operating, production, input, or implementation) objectives, which relate to the means (materials, leaders, facilities, etc.) used to operate the programme, the amount of energy expended in the programme delivery, or matters related to client participation. (Example: to increase the awareness of endangered plants.) The second type are performance objectives. These are the measurable end products of the programme, that link directly with the mission and goal statements. (Example: participants will have identified at least one half of all bird species on the identification list.)

# Needs and assets

An understanding of (1) the needs of participants and (2) resource assets, is critical both for the decision to start a new ecotourism programme and for the choice to continue with a programme after it has been offered. Needs and assets are not static, but rather change over time. Consequently, a periodic assessment of these helps to keep the programme timely and efficient. The ecotourism service provider will be confronted with a pool of needs (see Knowles 1980), which derive from participants, the organisation itself, and the community. All of these will need to be filtered according to the overall requirements of the ecotourism operation, as well as the various sources of need. Based on the philosophical orientation of the service provider, such needs will be operationalised through a series of programme priorities, which will provide the basis of the ecotourism programme.

## Needs assessment

The first step, and one of the most important in the process of developing ecotourism programmes, is the identification of participant needs which the service provider will seek to isolate and later satisfy. Indeed, it should be the prime directive of service providers to strive to meet or exceed the expectations of participants through the programmes that they develop (McCarville 1993). To do this, the programmer is responsible for tapping into the needs of participants for the purpose of developing effective programmes and generating meaningful experiences. The vehicle which has been used to accomplish this end in the recreation field is the needs assessment, which incorporates any number of different survey or measurement tools to explore needs. (Tourism planners have often employed tourism strategies or tourism marketing plans which focus on aspects of demand from generating regions, as well as on the state of affairs within destination regions, including capacity, attractions, and impacts (Veal 1992). The extent to which these effectively address recreational or tourism need, in the manner addressed in this chapter, is open to debate.)

The needs assessment is 'the process by which the program planner identifies and measures gaps between what is and what ought to be' (Windsor *et al.* 1994: 63; see also Gilmore and Campbell 1996). Such needs may be identified for an individual, but more typically for groups of people (grouped by gender, age, geographic location, and so on). Table 7.2 addresses some of reasons why practitioners conduct needs assessments (adapted from DeGraaf *et al.* 1999). For example, under service orientation, the needs assessment may allow tourists to provide valuable input into the programme; it may allow for new ideas in creating innovative programming; it may allow the service provider to serve as many people as possible; and it helps the programmer stay in tune with psychological and physiological needs of participants. Depending on the time, situation, resources, and experience of the programmer, a number of steps will need to be followed in conducting the needs assessment, including the following general categories: (1) defining need, (2) identifying study subjects, (3) asking the right questions, (4) collecting the data, (5) analysing and interpreting data, and (6) reporting and using the data for planning.

## Inventories of attractions and resources

Just as the needs of participants change from year to year and from programme to programme, so too do the base of attractions and physical, natural, recreational and socio-cultural resources which form the basis of ecotours. In view of this, the service provider should regularly inventory attractions and resources in order to document these changes. In general, such an inventory is a systematic compilation of attractions and resources which occur within a tourism destination region, and which demarcate the spatial and temporal

**Table 7.2 *Why conduct needs assessments?***

| Reason | Description |
|---|---|
| A. Service Orientation | |
| Tourist input | Programming decisions may be influenced positively by allowing tourists the opportunity to share their views |
| Solicit new ideas | Allows for creativity and innovation in programme planning |
| Inclusion | Enables the provider to serve as many people as possible, regardless of age, sex, ability, race, religion, education |
| Meet real needs of people | Helps the programmer stay in tune with the physiological and psychological needs of participants |
| B. Desire for quality | |
| Professional commitment | Operators must strive to be ethically based and to exceed standards which may be set for the industry/region |
| Resource allocation | Helps to inform the operator on where s/he needs to focus (e.g., more staff, better transportation, and so on) |
| Accountability | Being responsible for our actions and who and what they impact (tourists, parks, local people, other land users) |
| Increase profits | Important element to ensure that the company makes profits while ensuring tourist satisfaction |
| C. Programme Management | |
| Manage duplication | Duplicating services within or between operations is a waste of resources and leads to unhealthy competition |
| Address safety | How to better use facilities, transportation, and better understand hazards in very dynamic environments |
| Plan a variety of programmes | Help us design and implement the best possible programmes for the specific targets that we cater to |
| Develop/meet individual goals | To help providers and participants to meet their goals, such as skill development and environmental education |

Source: Adapted from De Graaf *et al.* (1999)

boundaries of the ecotour. It is the task of the operator not only to list these potential attractions and resources but also to identify any special significant factors related to these attractions which would allow the operator to generate greater value in programme offerings.

A comprehensive list of general attractions and resources is included in Table 7.3. The table includes the 'presence' or 'availability' of a resource, the 'quantity' of the resource (e.g., waterfalls), as well as the 'quantity' of the resource for the purposes of the ecotour. It also includes a means by which to assess subjectively the ability of the resource to absorb the demands of a tour, which could include selected composite measures such as temporal and spatial aspects of the tour, levels of impact, levels of use, transportation, and so on. The operator may also record a subjective assessment of the future potential of the resource.

Once completed, the information yielded from this process will be an essential component of the programming planning stage. Also, the operator should search for individuals or groups – human resources – in the community who may be able to help in the interpretation of the environment. Every community has people who are knowledgeable about the natural and human history of the region, and who may be willing to share this information with ecotourists. For this reason, the identification of a number of specialists (e.g., through local

**Table 7.3** *General resource inventory*

| Resource | Presence/ availability | Quantity (no.) | Present quality (rank 1–10) | Capacity to absorb tour (low, med., high) | Future potential (low, med., high) and notes |
|---|---|---|---|---|---|
| **Natural attractions** | | | | | |
| Beaches | No | – | – | – | Nothing available, not appropriate |
| Birding | Yes (etc.) | Many types | 8 | (medium) | Migratory and resident species (high) |
| Karst landscapes | | | | | |
| Temperature | | | | | |
| Rainfall | | | | | |
| Average sunshine | | | | | |
| Desert environment | | | | | |
| Forests | | | | | |
| Fungus | | | | | |
| Geological formations | | | | | |
| Hiking trails | | | | | |
| Islands | | | | | |
| Lakes | | | | | |
| Lichen | | | | | |
| Mammals | | | | | |
| Mountains | | | | | |
| Nature trails | | | | | |
| Parks/protected areas | | | | | |
| Plants | | | | | |
| Rivers | | | | | |
| Waterfalls | | | | | |
| Wilderness areas | | | | | |
| Cultural/historical attractions | | | | | |
| Accommodation | | | | | |
| Transportation | | | | | |
| Food and beverage etc. | | | | | |

*Source:* Adapted from Fennell (2002b)

conservation groups) will help put variety in the programmes, with the added benefit of building strong support in the community.

## Programme design: structure

The programme structure involves five main components, each designed to provide an initial programme shell from which to build upon further. These include:

1   *Programme areas*. Ecotourism programme areas are activities which form the basis

of the actual programme. From a purely recreational perspective, and at a broader level, these activities include hobbies, music, outdoor recreation, mental and literary recreation, social recreation, arts and crafts, dance, drama and sports and games. Ecotourism programme areas would include, for example, bear viewing, whale-watching, stargazing, participation in wolf howls, birding, and perhaps other more marginal ecotourism activities like canoeing and kayaking – depending on the motivations of the participant.

2   *Programme formats*. Ecotourism experience will be structured on the basis of one of many different programme formats (see Table 7.4). Obviously a programme format such as 'trips and outings' will be most often used by ecotourism service providers. This is not to say, however, that it is the only one, or that more than one format could not be used for a programme. Ecotourism providers may develop 'instructional' classes for their potential clients, including, for example, slide shows on the birds of Costa Rica. This breakdown of different formats suggests that while certain activities are best presented using one type of structure over another, it is also important to realise that the appropriate format will enable guides and staff to perform at the levels to which they feel most comfortable. The following are formats identified by DeGraaf *et al.* (1999), and include those which have been developed from a conventional recreation context: club and special interest groups, competitions, trips and outings, special events, instructional, drop-in, service opportunities, and outreach.

**Table 7.4 *Programme design matrix***

| Programme area | Programme format | Programme setting | Programme mobility | Programme lodging |
|---|---|---|---|---|
| *Nature oriented* | | *Backcountry* | *Motorised* | *Non-fixed Roof* |
| Bird census | Club | Marine | Cruise liners | Hammocks |
| Nature games | | Freshwater | Powerboats | Tents |
| Erosion control | Competition | Terrestrial | Aeroplanes | Caves |
|   projects | | Aerial | Helicopters | Open shelters |
| Nature crafts | Trips | Combined | Trains |   (lean-to) |
| Wolf howls | |   settings | Cars and buses | Tepees |
| Tree/plant | Special events | | Off-road vehicles | Igloos/snow |
|   identification | | *Rural* | Snow vehicles |   caves |
| Stargazing | Instructional | Marine | | Bivouacs |
| Nature walks | | Freshwater | *Non-motorised* | |
| Wildlife viewing | Drop-in | Terrestrial | Canoes and | *Fixed-roof* |
| Hang gliding | | Aerial |   dugouts | Hotels and |
| Outdoor | Service | Combined | Kayaks |   motels |
|   photography | |   settings | Rafts | Resorts |
| Spelunking | Outreach | | Yachts | Bed and |
| Snorkelling | | *Urban* | Gliders |   breakfasts/Inns |
| | | Marine | Balloons | RV/caravan parks |
| *Adventure* | | Freshwater | Mountain bikes | Ecolodges |
| *Culture* | | Terrestrial | Skis and | Ranches |
| *Farm and ranch* | | Aerial |   snowshoes | Huts and cabins |
| *Heritage* | | Combined | Horse, llama, | Houseboats |
| (etc.) | |   settings |   camel | |
| | | | Hiking/ | |
| | | |   backpacking | |
| | | | Wheelchairs | |

*Source*: Adapted from Fennell (2002b)

3   *Programme setting.* Even though ecotourism has been a concept intricately tied to wilderness settings, more and more research is emerging to suggest that it is also an urban and rural phenomenon. As such, these three settings may be matched with at least five sub-settings, including marine, freshwater, terrestrial, aerial, and a combination of these (see Table 7.4).

4   *Programme mobility.* Forms of transportation may be categorised according to motorisation and non-motorisation. In the case of the former, this may include cruise liners, powerboats and helicopters; in the latter, canoes, rafts, balloons and snowshoes (see Table 7.4).

5   *Programme lodging.* The success of ecotourism experiences is hinged, in part, on the type of accommodation built into the programme. These may range according to units which have a non-fixed roof, and those with a fixed roof. Non-fixed roof units include hammocks, tents, caves, tepees, lean-to's, igloos, and bivouacs; fixed-roof dwellings include hotels, motels, resorts, bed and breakfasts, recreational vehicles, ecolodges, ranches, huts, houseboats and tree houses.

Once programme areas, formats, setting, mobility and lodging options are selected they may be fitted within a broader programme matrix. This allows for an appreciation of the magnitude of different options available to the service provider, while at the same time providing the basic structure of the programme (essentially the shell of the programme). Programmers select the most appropriate options for the programme within each of the five categories, on the basis of availability, what might appeal most to clients, or other issues including the goals and objectives of the programme and the strength and weaknesses of staff and available resources. This may entail a great deal of brainstorming on the part of programme staff, which also may include the development of several programming scenarios. These may be weighted or judged according to any number of criteria, including time, budget, equipment, human resources, risk, and so on.

## Programme design: logistical considerations

Of the many aspects of programme planning, this section contains a description of various logistical and practical aspects which are important to the physical and mental health of the participants. This starts with the briefing of clients, whether this be a month ahead of the trip or the day of the trip, depending on the ecotour. This information can be rather specific to the tour, as regards animal checklists and so on, or it may be of a general nature, with respect to passports and visas, vaccinations, health insurance, importing goods, luggage, money, and home security. In briefing ecotourists, it is also helpful to rate tours on the basis of terrain type, length, and physical exertion required. This provides the needed information enabling ecotourists to place themselves in the tour which best suits them physically. Attracting individuals to a tour who are not mentally or physically prepared for the rigours of the trail accomplishes nothing, and in fact, depending on the risk involved, might place one or many individuals in danger.

## Reasons to be cautious

As outlined above, there are many reasons why ecotour operators need to be cautious in the development of their tours. It is widely recognised among those who programme in the field of recreation that the most hazardous part of a programme is not the activity participation phase but rather the transportation of participants from one site to another (Gilchrist 1998). To ensure that accidents do not occur, service providers must be cautious in who they enlist to drive vehicles, the vehicles themselves which are chosen to transport, and the length of time that drivers are asked to transport participants.

No matter how civilised or primitive the programme setting, there must be the utmost concern for the manner in which food is stored, served and disposed of. It also follows that to keep food safe is to know the environment in which one is travelling. Urban environments will be different than wilderness ones, while hot and humid environments will be different than cold and dry ones. Furthermore, it is not just bacteria and protozoans which need to be considered in keeping food safe, but also ants, mice, raccoons and bears. In wilderness settings, operators may need to use airtight containers for food storage, or they may opt to keep food safe by hanging it from trees and thus out of the reach of nocturnal animals. It should also be noted that guides must be open about cleanliness, by making it a practice to wash hands and dishes often and in full view of participants, to ensure that both parties feel comfortable with the food-handling process.

Stomach ailments appear to be one of the most common health problems for travellers. The usual culprits are food and the consumption of contaminated water. In the case of the latter, contamination comes in the form of bacteria (e.g., *E. coli*, *Salmonella typhi*, *Vibrio cholerae*), protozoans (e.g., *Giardia lamblia*, *Cryptosporidium*) or viruses (e.g., hepatitis), which appear to be much more prevalent in tropical destinations (Brock 1979). The largest of these, the protozoans, have been the topic of great interest, especially in relation to the devices which have been designed to eliminate them from drinking-water supplies. Because of their size (10–100 micrometres in size, compared to bacteria which are generally 2 micrometers in size, and viruses which are 0.1 to 0.2 micrometres), protozoans are more easily captured through the various water filtration devices that are on the market. Pathogens in water can be eliminated in one of three ways: boiling, chemicals such as iodine, and filtration. The operator should keep in mind that although generally effective, protozoans, like *Cryptosporidium*, are resistant to iodine, and some people react negatively to this substance.

Operators should insist that their guides carry up-to-date backcountry first-aid guides in the event that first-aid is required. There should be the recognition that first-aid in wilderness settings is different than first-aid in urban settings. In the latter case, responsibilities surround breathing, bleeding, splinting and safety. In the wilderness, care goes beyond what is presumed to be 'first-aid' to other concerns related to exposure, shelter, travel, sanitation, innovation, and the administering of proper fluids and food (Merry 1994). Furthermore, guides should be aware of access and egress points while in the field, know where the nearest healthcare facilities are located and how to get there, carry a cellular phone, as well as a list of a number of pertinent contact numbers in the event of an emergency. In addition, guides should be encouraged to seek advanced certification in first-aid in order to be as current, and as prepared, as possible.

The range of health risks that travellers are exposed to is extensive, considering the number of plants and animals which have the ability to inflict pain or discomfort. Concern usually exists over the largest and most toxic animals which inhabit a particular region (e.g., crocodiles, spiders, and snakes in Australia, bears in Canada, snakes in India). In reality, it is the microscopic organisms, as we have seen above, which usually cause most of the problems. The effects of plants and animals may be indirect by, for example, dissuading people from travelling to a region, or direct, such as in the case of a first-hand encounter with a plant or animal. For example, mites and ticks are carriers of pathogens which are the cause of many human diseases such as Lyme disease, Rocky Mountain spotted fever, and tularemia. Lyme disease, which is especially well known in North America, is a case in point, where the secretions from the deer tick, *Ixodes dammini*, infect human hosts with the bacterium *Borrelia burgdorferi* through its feeding behaviour. The deer tick has highly specialised hooked mouth parts that enable the invertebrate to maintain a solid hold of the host. To aid in its feeding, the tick will also flatten itself on the host to avoid being brushed off. Early symptoms of the disease include a skin rash at the point of contact, and later a doughnut-shaped rash. This is accompanied by flu-like symptoms, swollen glands, stiff neck, backache, nausea and vomiting. Later stages of the illness may include problems with joints, inflammation of the heart, neurological problems (temporary paralysis of face muscles), with many of these symptoms highly variable between patients.

## Programme design: risk and leadership

Texts on recreation programme planning typically include leadership as an important aspect of the planning cycle at this stage or perhaps the following stage. For the purposes of this discussion, it has been included in the previous chapter to fit with the discussion on professionalism. The author has elected to emphasise the importance of risk management at this stage of planning, which is thought to be a logical next step from the discussion in the previous section.

## Risk management

At its most fundamental level, risk management is the process developed by an agency to avoid costs. Generally, these costs can be as a result of injury, damage to equipment and facilities, as well as wrongful actions. The main sources of risk (i.e., the things that exist in one's programme which contribute to the types of risk outlined previously) include facilities, equipment, the programme itself, and people (Corbett and Findlay 1998). More succinctly, risk management may be defined as the 'formal process of assessing exposure to risk and taking whatever action is necessary to minimize its impact' (National Association of Independent Schools, cited in Ammon 1997: 174). It is thus a sophisticated phrase for a very basic management tool, and the need for an appropriate risk management plan cannot be understated for all recreation and tourism providers. Naturally, however, in consideration of all these various forms of tourism and recreation, some, like adventure pursuits, require very detailed and specific plans, for, as Liddle (1997) suggests, such activities are based upon the perceived potential for loss, with the occurrence of actual loss – which may mean the loss of life for these activities – being unacceptable. In general, the risk management plan will involve a series of steps which will enable the operator to identify risks, classify these, develop risk control measures, implement these measures, and monitor and modify (Watson 1996). The identification and classification of risks may be accomplished by outlining a number of steps to enable the programmer to identify problematic areas for each and every programme, as outlined in Table 7.5.

Potential risks are then subject to an evaluation of either 'low, medium, or high', or perhaps some other method such as '1' (low) to '10' (high). For example, under the category of 'location' the site may be evaluated as low, the weather as low, the routes and accommodation as medium, and the facilities as high. Corbett and Findlay (1998) suggest that the next step is to determine the measures which may be employed to control the significant risks which have been identified on the basis of the following four options:

1 *Retain the risk*. Nothing is done because the likelihood of occurrence is low and/or the consequences are slight.
2 *Reduce the risk*. Practical steps are taken to reduce the likelihood of occurrence and/or the consequences, usually by changing human behaviour or actions. This may include designing and following regular maintenance programmes for equipment and facilities, helping staff to pursue appropriate professional development, or posting appropriately worded warnings and rules within a facility.

**Table 7.5** *Outline for a risk-management plan*

| | |
|---|---|
| 1 General description | 6 Participants |
|   (a) Name of program |   (a) Number |
|   (b) Type of activity |   (b) Skill level |
|   (c) Level |   (c) Characteristics |
| 2 Dates and times | 7 Leaders |
|   (a) Dates |   (a) Number/roles |
|   (b) Times |   (b) Qualifications |
| 3 Goals and objectives | 8 Equipment |
|   (a) Organisational |   (a) Type and amount |
|   (b) Activity |   (b) Control |
| 4 Location | 9 Conducting the Activity |
|   (a) Site/area |   (a) Preactivity preparation |
|   (b) Weather |   (b) Group control |
|   (c) Routes/accommodation |   (c) Teaching strategy |
|   (d) Facilities |   (d) Time management |
| 5 Transportation | 10 Emergency preparedness |
|   (a) Mode |   (a) Policies |
|   (b) Routes/destinations |   (b) Health forms |
| |   (c) Telephone numbers |

*Source*: Adapted from Ford and Blanchard (1993)

3 *Transfer of risk.* The level of risk is accepted, but passed on to others through contracts. This may include the insistence that all participants sign a waiver of liability, the purchase of insurance which is appropriate in scope and amount for all activities, or the contracting out to other agencies (e.g., transportation).

4 *Avoid the risk.* Steps are taken to restrict, limit, postpone or cancel certain activities. This may include postponing events in dangerous conditions, not travelling in bad weather, restriction to participants with certain levels of expertise, insistence on maintaining a 'no alcohol' consumption policy for certain activities, or adhering to an organisation's or region's policies.

Bodansky (1994) observes that risk is a function of the magnitude and probability of harm. It involves estimates of probabilities about whether certain events will occur or not. In adventure recreation, risk is normally understood in two ways. Real risk is the true potential for loss, where no loss means zero risk, and death relates to extreme risk (Priest 1990). Real risk cannot be estimated with certainty at any one time. On the other hand, perceived risk amounts to the best estimation of real risk. Those who have gained extensive experience at high-risk activities such as mountain climbing can better understand the inherent risk of an activity. The options outlined above are structured in order to provide enough flexibility to service providers in helping to make decisions which are not only in the best interests of the provider (as regards cost and resource accountability, not to mention liability), but also to the ecotourists who may participate in an activity knowing that the operator has done as much as possible to control for the expected and unexpected.

# Programme implementation

The effort placed into planning will have merit if it leads to programme implementation. This step of the programming process can include marketing, quality, staff training, public relations, budgeting, implementation strategies, and schedules and itineraries. For the purposes of this section, only implementation strategies will be considered.

McKenzie and Smeltzer (1997) argue that programme implementation can occur successfully if consideration is given to: (1) the resources that are available for the programme (as suggested in the section on budgeting, and (2) the setting for which the programme is intended – which is especially relevant to ecotourism. Programme implementation, therefore, may involve a significant outlay of resources, including time, personnel and capital. These resources are best viewed in the context of a specific strategy.

A model discussed by McKenzie and Smeltzer, developed by Borg and Gall (1989), is based on a series of field testing and revision stages which allow the programmer to secure a heightened level of programme acceptance and momentum. Its strength lies in the ability to change the programme at various stages as well as to drop the programme altogether without a significant outlay of capital if, for example, participants are greatly dissatisfied with it after step 3.

This implementation strategy includes an element of programme building, which is strongly recommended by McKenzie and Smeltzer. While it appears to be about as thorough as one needs to be in the construction of a programme, the authors suggest that the process may be modified or scaled down to suit the needs of the service provider. The importance of the programme relative to other programmes offered, in light of time and other resources, will be an important consideration in assessing the level of detail required for this approach. Having said this, it should be realised that those who follow through with the model in all likelihood stand a greater chance of developing more successful programmes later on.

**Table 7.6** *Major steps of Borg and Gall's research and development cycle adapted to programme planning*

1 *Preliminary field testing.* Programme given to a few individuals from the target population, with interview, observation, and survey data collected and analysed.

2 *Main programme revision.* Based on results of preliminary field test.

3 *Main field testing.* Programme given to about twice as many individuals as in step 1. Pre- and post-programme quantitative data collected; results examined with respect to programme objectives.

4 *Operational programme revision.* Based on results of the main field test.

5 *Operational field testing.* Programme given to about twice as many clients as step 3. Interview, observation, and survey data collected and analysed.

6 *Final programme revision.* Based on results of the operational field test.

7 *Dissemination and implementation.* Programme shared with others and implemented.

*Original source*: Borg and Gall (1989); this table adapted from McKenzie and Smeltzer (1997).

# Evaluation

While evaluation is often thought of as a process which is employed at the end of a programme only, it is actually an ongoing process which enables service providers to control the events that go on within the programme setting. Its importance to the overall programme planning process is underscored by Hall and McArthur (1998), who suggest that organisations which fail to evaluate are not committed to (1) understanding performance, (2) learning from mistakes, and (3) self-improvement. Evaluation can be defined as 'the identification, clarification, and application of defensible criteria to determine an evaluation object's value (worth or merit), quality, utility, effectiveness, or significance in relation to those criteria' (Worthen *et al.* 1997: 5). In a leisure context, evaluations are undertaken by programmers to answer some of the questions about what they had hoped to be able to accomplish. These questions often surface in the form of predetermined goals, discussed earlier in this chapter, where effectiveness is ascertained through a process involving any number of qualitative and quantitative techniques. This information arms the service provider with the ability to demonstrate accountability with regard to finances, demand, benefits, or the wise use of resources (Busser 1990). Evaluation can take place in the following areas: (1) administration (e.g., planning, organising, staffing, training, directing work, exerting control and inspiring creativity), (2) leadership or personnel (the qualifications, competence, and performance of staff), (3) programme content (meeting needs of participant, scheduling, transitions, goals and objectives, supervision, records, and so on), and (4) physical properties (including design, construction and maintenance of facilities and equipment).

## Formative and summative evaluation

Based on the previous discussion, evaluation should be employed at stages where it is thought to be most beneficial, including programme development and implementation stages. Evaluation at earlier stages is referred to as formative evaluation, and focuses on how best to initiate the programme. For example, it may be used to fine tune the informative aspect of a wolf-howling programme at selected points of the survey. Formative evaluation provides immediate feedback on the value of the interpretive tools, which may thus be built into the programme at later dates.

Summative evaluation provides programme decision-makers and consumers with judgements about the programme's predetermined merit in relation to criteria. Using the example of the wolf howl programme above, a summative evaluation would seek to determine the degree to which the curriculum improved the quality of the programme and the level of satisfaction of ecotourists, and thus the future value of the programme. The element of continuation is important, because it is a key concept in differentiating between formative and summative evaluations. While formative evaluation leads to decisions about programme development and implementation (modification and revision), summative leads to decisions about programme continuation, termination, expansion, or adoption (Worthen *et al.* 1997).

## How to evaluate

Evaluation has been formally organised into a number of techniques which may be used, depending on what needs to be evaluated. Techniques range from intuitive evaluation, which is a non-systematic 'gut-reaction' approach that allows for changes on a day-to-day basis, to more sophisticated approaches such as the goals-attainment scale which involves identification of goals and objectives and then an assessment to determine the discrepancy between the expected performance/outcome and the actual or realised performance/outcome. A popular evaluative technique which is especially relevant to ecotourism programmes is the Evaluation by Standards approach. In this form of evaluation, professionally developed standards, often developed by respected professionals in the field, are established to act as levels of achievement or goals (e.g., for safety purposes) which must be satisfied (Kraus and Allen 1997). For example, rafting operators would have to ensure that participants wear helmets at all times while in the raft, and that participants must be able to swim 50 metres before being admitted to the programme. This method focuses more on administrative details than customer satisfaction, and the prescribed standards which are developed often reflect the values of the agency or professionals responsible for their development (e.g., Association for Experiential Education, which has established a whole series of standards for a number of outdoor recreation activities). The same holds true for accreditation. The province of Saskatchewan has recently developed a whole range of professional standards which prescribe exactly what operators must do to ensure an appropriate standard of care (e.g., considerations of design, ecolodge considerations, enhancing the nature experience) in order to gain accreditation (see Figure 7.3). Farrell and Lundegren (1993) write that this is one of the most common evaluation approaches used by those in the public domain. Other groups who develop standards include, for example, the American Camping Association, the Australian Ecotourism Accreditation Programme, and other agencies who maintain a vested interest in the proper and safe conduct of those working in the field.

Evaluation equips the operator with the required information to gauge the significance of the programme. With this information in hand, the service provider has what is needed to decide the fate of the programme on the basis of a number of different options, including: (1) continue the programme with no modification, (2) continue the programme with modifications, (3) end the programme and repeat it at some other time, (4) end the programme and modify it at a later date, or (5) end the programme altogether. Obviously in cases where the programme has been quite favourably evaluated, little or no modification is necessary (although like any true professional the service provider will naturally want to tinker with the programme to make it better). At the other end of the spectrum negative evaluations would mean that the programme, especially if it has been negatively viewed in the past, should be shelved or removed completely. However, as Carpenter and Howe (1985) quite rightly suggest, there needs to be a very tight relationship between programme revision and needs assessment (the beginning of the next programme cycle). Without assessing the participants' needs from time to time, programmers will be unable to know if the demand for a terminated programme can be rekindled. The link between the evaluation–needs assessment cycle is effectively articulated by these authors, who write that 'Program revision allows you to be immediately responsive to your clientele as a result of your evaluation findings, while needs assessment keeps you informed of the changing demands for leisure programs and services, enabling you to remain responsive over the long run' (p. 172).

| | | | |
|---|---|---|---|
| ❏ yes | ❏ no | ❏ n/a | Does the operation include accommodation that directly involves and enhances the activities? |
| ❏ yes | ❏ no | ❏ n/a | If portable accommodation facilities (tents/tepee) are used, are they designed for minimum impact? |
| ❏ yes | ❏ no | ❏ n/a | Does the accommodation enhance the 'nature' experience? |

**Figure 7.3** Accreditation standards for sustainable accommodation

Source: Ecotourism Society of Saskatchewan (2000)

---

**CASE STUDY 7.1**

*Wilderness tourism and outfitting in the Yukon, Canada*

The interpretation of ecotourism takes a more general form in the Yukon Territory of Canada. The territory's selection of 'wilderness tourism' as an umbrella term for a variety of different back-country experiences is very much a function of geography (the remote wilderness setting of the territory) and the existing products and resources operating in the region. The undeveloped wilderness setting is thus a key factor in the delivery of tourism products in the region. Wilderness tourism is defined as 'tourism related to nature, adventure, and culture, which takes place in the "backcountry" and is primarily associated with multi-day trips, although it also includes day trips' (Tompkins 1996). Wilderness tourism is divided into (1) hunting, (2) fishing, and (3) adventure. This latter category is very much a reflection of the adventure tourism definition and activities of the Canadian Tourism Commission. Such activities include nature and wildlife observation, land adventure products, water adventure products, winter adventure products, air adventure products, and native tourism.

---

Programme demise or death for those which are unpopular may simply mean that there may be a time in the future when the programme is able to generate enthusiasm.

## Conclusion

At the outset of this chapter it was suggested that there are many commonalities between tourism and recreation, especially from the experiential standpoint. Service is a key aspect which is common to both areas, but there does not appear to be the same level of importance given to service in tourism, as there is in the field of recreation, where the delivery service system is designed to ensure that the participant is satisfied. Perhaps because of the relatively new focus on tourism in municipalities and other regions, there is a preoccupation to view tourism as nothing more than a vehicle by which to make money – to support businesses in the area. The tourism industry, however, will need to adapt if the services we provide for tourists – as participants, not clients – are to be viewed as something more than cash cows. More and more, the sophistication and the competitiveness of tourism services demands a different approach to service provision. We can go so far as to suggest that, all things considered equal, the agency which is a more proficient programmer will stand a greater chance for financial success. In doing so, however, the by-product is a participant base which will have been satisfied in any number of different ways. In the twenty-first century, tourists want to live their travel experiences in the same way they live their recreation experiences. This means education, skill development, exercise and self-actualisation. It also means that travel will need to be programmed in the same way that recreation experiences have been for years.

## Summary questions

1   In general, how can needs assessments and assets inventories allow ecotourism operators to help themselves, and, in doing so, benefit ecotourists?

2    How is it that programme planning, as discussed in the chapter, might enable the ecotourism service provider to value other things beyond the profit motive? Will such an emphasis ultimately help the provider remain competitive in the marketplace in the long run?

3    Provide a general overview of the programming planning model. Why is the cyclical nature of this type of planning so important?

4    How do formative and summative evaluation techniques differ?

5    State some of the professional responsibilities that ecotour guides have in the development of ethical ecotourism practices.

# 8 Ecotourism development: international, community, and site perspectives

This chapter focuses on tourism development from three perspectives: international tourism development, including development theory, tourism in underdeveloped countries, and the core–periphery concept; community development; and site development. While the international development literature provides a clear indication of where we have been in terms of tourism development, the community development or community-based ecotourism initiatives provide a contrasting view in support of where ecotourism should go in the future. The ecolodge phenomenon is included as an example of how site development can occur using sustainable development design principles and local decision-making.

## International issues

### Development theory

> I sit on a man's back, choking him and making him carry me, and yet assure myself and others that I am very sorry for him and wish to ease his lot by any means possible, except getting off his back. – Tolstoy

Development theory explores the economic and psychological link that exists between rich and poor countries. The relationship continues to be one that often contrasts meaning and value with respect to the use of natural resources and capital. Tourism is a sector of the world's economy that has tended to underscore and exacerbate the inequalities that exist between core metropolitan nations and peripheral, marginalised economies, and in doing so perpetuates the division between those who have access to resources and those who do not. To Vogeler and DeSouza (1980), this division has conjured up images of distorted human beings and grotesque folklorism, where the underdeveloped are like children unable to fend for themselves and with little capacity to make decisions.

Dependent countries were not considered economic entities until the late 1890s, when a new, 'lofty' responsibility regarding dependent peoples emerged (British Africa encouraged participation in the cash economy). This lofty purpose was coupled with a crude, hypocritical view of natives and administration, with the recognition that external power had a responsibility to develop the resources of dependent countries for the benefit of the whole world. Some saw the acquisition of overseas dominions as motivated by strategic reasons, while others sought only economic incentives (Brookfield 1975).

In measuring development, past research has used a variety of means to define it, including number of hospital beds, daily intake of calories, daily intake of animal protein, percentage of population in agriculture, food as a percentage of total imports, export economy, infant mortality, death rate, and aid per capita. Many of these have been criticised owing to the fact that the socio-cultural, environment and economic landscapes of each country are different. For example, food requirements are a cultural phenomenon, where countless generations have formed the habits of food intake seen today. Comparing the daily intake of calories between North Americans and Southeast Asians is pointless. Their physiological requirements are quite different (see Mountjoy 1971).

Resources have also been used to gauge the development status of nations, where an abundance of resources is thought to be an indicator of development. However, Mountjoy (1971) points out that even though Brazil has an abundance of resources and Denmark few, Denmark is considerably the more developed. Thus, development should be based not upon potential, but on existing levels of material welfare. Switzerland is a further case in point, where capital is a substitute for natural resource endowment. The progression of development, according to Brookfield (1975), occurred as a result of capitalist countries broadening or diversifying production into new sectors. This was to be achieved internationally through a widening of resource exploitation of less developed countries (LDCs). Brookfield maintains that the most aggressive industrial and trading nations (e.g. Britain) benefited most. They put pressure on other countries to open their trade. Adam Smith's *laissez-faire* philosophy was the true avenue of growth unfettered; it would work towards optimal production.

In recent years there has occurred a major paradigmatic revision in development thinking that is presenting a fundamental challenge to the conventional consensus on economic development. This new approach emphasises the basic needs of the poor, and advocates a sensitivity for development at the ground level. Real improvement cannot, therefore, occur in LDCs unless the strategies that are being formulated and implemented are environmentally sustainable (Barbier 1987). Brookfield (1975) emphasises support of this contention: the study of development joins directly with the study of all change in human use of the environment, and provides elements of empirical evidence for infusion with other theories in the task of generating a dynamic human–environment paradigm.

Mountjoy (1971) stipulates that the most important factors in gauging development are people; their number, age, enterprise, initiative, inventiveness, knowledge and willingness to sacrifice. It is not the responses themselves (which are passive), but the way in which they are to be utilised that is deterministic of development. The Western nations' concern for invention is not shared by all societies. Technologically inferior societies exploit single landscapes, within the framework of their social structures, while the Westerner is said to exploit all environments (ibid.). This is most clearly emphasised by the recent international debates on the implications of global warming for the future viability of the planet. LDCs make a strong case against the more industrialised developed nations for their role in the creation of the pollution (e.g. nitrogen oxide and sulphur dioxide emissions) that has contributed to global warming and ozone depletion.

## Tourism in the underdeveloped world

LDCs suffer from a history of colonial domination and what has been referred to as dependent development. A review of tourism literature reinforces the notion that there is a set evolutionary pattern of multinational interests in the tourism sector of the underdeveloped. Dependency, therefore, can be conceptualised as a process of historical conditioning which alters the internal functioning of economic and social subsystems within a less developed country (Britton 1982). This dependency occurs not from processes within the LDCs economy, but rather from demand from overseas tourists and new foreign company investment in the LDC. For example, after a potential tourist destination has been identified (on the basis of unique biophysical or cultural conditions), the involvement of the multinationals increases. Foreign companies greatly influence the image of a destination country through development and promotion. Such efforts lead tourists to perceive the host country in terms of this image and the nature of hotel accommodation, attractions, and other tourist services as publicised.

Domination of the tourism sector in an LDC is most outwardly represented by the foreign ownership of airlines and hotels. However, the grip on the destination region often goes much deeper. Because of the inability of agricultural and manufacturing producers in most underdeveloped economies to guarantee a good-quality supply of goods and services for international luxury-standard tourist facilities, there is a strong reliance on imported supplies for both the construction and the operation of tourist facilities. Also, as Winpenny (1982) writes, middle and senior management levels of tourism developments in underdeveloped countries are often occupied by expatriates. The following scenario of metropolitan dominance serves as an example:

> On arrival in a Caribbean island between plane and hotel, the tourists could well have passed through a terminal building presented by the people of Canada and have been driven along a road either built or improved 'thanks' to Canadian aid, on behalf of the expected Canadian tourists of course. At the hotel you will likely be greeted with rum punches, the first of the few local products to be encompassed by the package. At your first meal, a menu, perhaps designed in Toronto, Chicago or Miami, will provide you with a selection of food imported for the greater part from North America – good familiar, homogenized, taste-free food, dressed up with a touch of local colour.
>
> (Hills and Lundgren 1977: 257)

The question therefore is as follows: who owns and controls the various components of a holiday package? It has been found (Economic and Social Commission for Asia and the Pacific 1978: 40) that where tour packages consist of a foreign air carrier, but include local hotel and other group services, destination countries receive on average only 40–45 per cent of the inclusive tour retail prices paid by the tourists in their home country. If both the airline and hotels are owned by foreign companies, a mere 22–25 per cent of the retail tour price stays in the destination country (see Chapter 5 for a discussion of leakages within an economy).

While Rajotte (1980) feels that tourism is less environmentally destructive than other forms of development that exist in tropical islands, there are still many significant effects that may be attributed to tourism. Many of these impacts are universal, but their intensity and severity are more noticeable in island environments (much of the past research on dependency and development has been linked to island environments). The notion that ecological diversity often comes with size is important. Nation-states of the Caribbean, because of their small size, have tenuous floral and faunal vitality. Tourism can stress these systems through yearly visitation levels that exceed the local population of an island; for example, Hayward et al. (1981) write that in 1979 more than 600,000 tourists visited Bermuda, some ten times the population of the island.

A key area of research for the future is the need to document whether or not developing countries such as Brazil, Costa Rica, Dominica, the countries of Eastern Africa, and Ecuador undergo the same social, ecological and economic dysfunctions from ecotourism, as outlined above, from the perspective of mass tourism. For example, Kusler (1991) writes that ecotourism in the LDCs is marked by limited dollars, government-directed financing, foreign visitation, foreign ownership of hotels and other facilities, and non-existent land use planning. In this sense, ecotourism as an intervention may or may not improve the socio-economic conditions of a country from the example of mass tourism outlined above. It is important to realise, as outlined by Kusler, that the LDCs and their developed-country counterparts are separated by a number of key structural differences. These include finances, the political climate of a country, accessibility to opportunities, discretionary income, and resources for the acquisition and planning of land.

## Core–periphery concept

One of the main theoretical approaches used to explain development is the concept of core and periphery, which Freidmann and Alonso (1974) describe as characterised by a dynamic, growing central region, and a slower-growing or stagnating periphery. The core is marked by high-growth potential whereas peripheral areas are often marked by declining rural economies with low agricultural production (loss of primary resource). Shields (1991: 3) refers to peripheralised regions as marginal places, regions that 'are not necessarily on geographical peripheries but, first and foremost, they have been placed on the periphery of cultural systems of space in which places are ranked relative to each other'. Important in his discussion is the fact that margins, from social, political, economic, and perceptual perspectives, are systems of centres and peripheries and are, therefore, locationally oriented. These spatial systems are established in a series of binary relationships (economic, society, and so on) of centre and periphery, so that margins signify everything that the centre denies or represses (Shields 1991). Our socio-economic paradigm is constantly exercising this binary relationship, and it reaches to the furthest outposts of physical space and humanity.

In a touristic sense, a classic example of the core–periphery idea can be found in Christaller's (1963) analysis of tourism location in Europe (see Chapter 2 and the discussion on the evolving nature of tourists within destinations). Christaller wrote that tourism avoids central places and is drawn to the periphery, reaching the natural resource base not found in cities. This idea has been further developed by a number of authors (Butler 1980; Pearce 1989). In particular, Battisti (1982) reflected on the principles established by Christaller on tourist space organisation, noting that tourism is part of the continual process carried out by humankind in order to specialise and to diversify the exploitation of the soil. Battisti made reference to the distinction between peripheral regions and peripheral places. The first represents a wide range of possibilities in choosing a recreational place. The second indicates real destinations of tourist flows, where the 'fruition of space for recreational purposes is accomplished' (ibid.: 60).

The core–periphery concept has also been applied more recently to the analysis of adventure travel regions from a spatial context. Zurick (1992) suggested that the movement of adventure travellers in Nepal takes place through a complicated hierarchy of gateways. Tourists from the core move through an international gateway (semi-periphery), to a national gateway (periphery), and further to a regional gateway (periphery frontier), said to be the adventure region. In this latter region there is often a clash between traditional interests and national development goals. As frontier regions succumb to further intrusions, the uniqueness of these areas diminishes, as does the potential for travel to untouched areas in the future as these areas become fewer in number.

One of the key defining elements of travellers to outward peripheries is the need to experience something different in a remote setting. Along with this, perhaps, is the need to escape the mainstream of tourism, which can only be achieved by taking the road less travelled. Although there are few data at hand to support this idea, the travellers mindset may alter upon reaching successive outward peripheries of a destination. Fennell (1996) thought that the following factors contributed to this feeling:

- *Familiarity with the destination.* A lack of knowledge or familiarity with a region's environment/culture may add to the uniqueness of such an area and therefore represent a perceived periphery.

- *Unscheduled change.* Movement that occurs spontaneously or that is not pre-planned (e.g. from a core region to a peripheral region).
- *Psychological and/or physiological change.* Personal adaptations that have to be made in order to feel comfortable or maximise feelings of satisfaction/pleasure, including culture, language, personal hygiene, food and shelter.
- *Distance from amenities.* This factor reinforces the notion that a particular peripheral region does not, cannot, or will not offer goods and services that can be found at regions closer to the core.
- *Adaptation.* As tourists move out to a new periphery, the accompanying shortage of goods and services such as shops and facilities – found at regions closer to the core – becomes more accepted. The state of the infrastructure, therefore, plays less of a role in satisfying the tourist. Poor weather may also not pose a problem because a more 'frontier-like' attitude is adopted.
- *Population density.* Regions with a lower tourist and local population density may provide an indication that the periphery has been, or is being, reached.
- *Authenticity.* As population density decreases, the authenticity and/or incidence of natural and cultural attractions may increase in relation to the characteristics of the region. The character of the region is important, as Costa Rica may have a higher species diversity in urban areas than Shetland has at the periphery.
- *Symbolism.* The periphery may be represented by experiential, environmental, or cultural phenomena (symbols). Such phenomena may include lochs, mountains, wind, solitude, barrenness, a rainforest, or lack of banking machines. Consequently, the more these symbols are experienced, the greater might be the feeling of reaching the perimeter. Fear may also act as a symbol, as may the realisation that there is a lack of knowledge of a region and its characteristics. Fear or anxiety of the unknown is an important feature of adventure tourism (characteristics that Csikszentmihalyi, 1990, has reported on regarding risk-taking in rock climbers).
- *Scale of attraction.* It is the region itself that may take on characteristics of 'attraction' in the outer periphery. For example, while cities have many attractions and goods and services to offer tourists and locals alike, small, peripheral islands may have comparatively few or no structural attractions, and because of this the nature of the island's limited infrastructure, small size, and distance from the mainland may be the attraction itself.
- *Distance from home.* This factor occurs in combination with other factors. For example, travel to Sydney, Australia, from rural Canada represents a large distance, but may not necessarily represent a mental change from core to periphery.

Thus the relationship between core and periphery has not only a psychological component but, as discussed earlier, a spatial one. This spatial component occurs at a variety of scales. Examples include small islands of the South controlled by the North, and huge territories of the North controlled politically, economically, and socially by decision-makers of the core. The Northwest Territories of Canada has been identified by Keller (1987) as an example of a peripheral region caught in transition regarding the development of its tourism industry. To Keller, differing authorities in control of development in this region become important actors in both decision-making and implementation of a viable tourist industry. These hierarchies of control occur at local, regional, national and international levels. As the sphere of tourism development becomes international, Keller suggests, the more the periphery has become 'a playground for exogenous investors, and the peripheral government an observer of its own fate' (ibid.: 25). It is stated by Keller that to avoid a transgression of the above scenario, the periphery (Northwest Territories) must act to control the decision-making in the development of a tourist industry, and limit the development to a scale of growth in tune with the resources, capital, manpower and culture within the periphery.

The implications of the core–periphery concept to ecotourism are that, often, remote travel regions are dependent on national and international markets, inbound and outbound tour operators, as well as externally based transport modes and schedules. With the inability and lack of resources fully to plan and implement ecotourism on their own, the cycle of tourism dependency continues for peripheral regions. This lack of an ability to link directly with the market in the developed world is one of the main reasons why the benefits of ecotourism do not percolate down to the community level; and it is precisely why those who are able to coordinate ecotourism activities – whether they be expatriates or local – between the LDCs and the developed world stand greatly to gain economically. In such a case the power to make ecotourism happen in a developing country lies in the hands of a few, and perpetuates the phenomenon of a typical LDC economic situation, with much of the money and resources lying in the hands of a few, and the majority of the population, in the absence of a prominent middle class, living in impoverished conditions.

While the preceding discussion has emphasised the significance and impact of externally based control in the development of tourism and ecotourism enterprises at a macro level, the discussion of community development to follow offers an alternative to many of the tourism development concerns outlined above. Community development, then, may be considered as a viable means by which to offset the conventional tourism development models of the past and redistribute control and decision-making among the individuals within the community, not to those from outside.

## Community development

Community development, as described by Smith (1990a), originated in the self-help programmes that were developed during the depression years in Canada, the United States and the United Kingdom. A defining characteristic of community development is that it is based on local initiatives, in that it advocates a site-specific approach to finding solutions to community problems using community members and community resources. Bujold (1995: 5) defines it as 'the process by which the efforts of the people themselves are united with those of governmental authorities to improve economic, social, and cultural conditions of communities'. Typical community development encompasses all aspects of the community and focuses on the best quality of life possible for its members (Davidson 1995), and may involve creating new business and employment, increasing cultural awareness, or providing a range of opportunities for all members of the community. However, Joppe (1996) alludes to the fact that it is important to understand that there exists a fundamental division between conventional community development and community economic development (CED) models. While conventional economic development focuses on the attraction of new businesses to the community (seen as an outward-directed approach to development), conversely a CED focus is on being small, green, and social, and is more inward in its orientation by striving to 'help consumers become producers, users become providers, and employees become owners of enterprise' (Joppe 1996: 476), through the principles of economic self-reliance, ecological sustainability, community control, meeting individual needs and building a community culture.

Tourism is increasingly seen as a key community development tool in the 1990s, with the recognition of its economic contribution in bolstering stagnating economies and diversifying existing sectors, and its ability to unify community members. Such is the case in the Shetland Islands, Scotland, where tourism is being relied upon to sustain an economy that was once dominated by North Sea oil development (Butler and Fennell 1994), or the Finnish island of Åland, where all tourism initiatives are owned or controlled by local people (Joppe 1996). With this realisation, researchers have begun to explore community development as it relates to tourism and ecotourism from many different perspectives. As an example, Christensen (1995) examines the role that tourism can play in securing quality of life within the community. At the foundation of any tourism development strategy is the realisation that

> If tourism development is to be viable as a long-term economic strategy, these concerns [social and ecological] must be addressed, and the resource base must be protected in the process. The host community is the economic, social, cultural, and infrastructural resource base for most tourism activity, and resident quality of life is a measure of the condition of the resource.
>
> (Christensen 1995: 63)

Christensen proposes a community quality of life framework that addresses both objective and subjective indicators at the individual and community scale. A tourism development project is said to effect changes in the quality of life of members of the community, which in turn cause impacts at different social scales. Such impacts need to be evaluated both by individuals and by the community, depending on the scale of reference to the individual. There is a deliberate hierarchy established with respect to agents of change, suggesting that appropriate solutions need to be implemented at the individual social scale first, followed by neighbourhood groups, coalitions, responsible industry, policy and governmental regulation.

The tactical approaches used by regions (local or national governing bodies) will make or break how ecotourism is perceived by local people, according to MacKinnon (1995). She cites a number of examples of communities in Mexico which have adopted or been identified to develop ecotourism in the country. One such community is Tres Garantías in the Yucatán peninsula. This region was initially set out as a hunting reserve by an international development team, but the area's residents found that non-consumptive recreation was more attractive to tourists than consumptive forms. The project, however, was conceived using a top-end approach, and even though there have been isolated benefits derived from ecotourism, MacKinnon suggests that social integration within the community is not complete, and will not be complete until benefits are more widespread and the perception of ownership is given to the community. MacKinnon further uses the example of Los Tuxtlas, Veracruz, to illustrate that ecotourism has been successfully integrated into a community through more of a grassroots approach (initiated through the efforts of a local woman who set out to build empowerment slowly and deliberately). In this latter case, the community has been more quickly able to realise benefits which have been spreading on a more regional scale. According to MacKinnon, the traditional mass tourism development model perpetuates the widening gap between rich and poor in countries like Mexico. Conversely, community-based tourism 'lends itself to being environmentally sustainable. Locals are the first to recognize the benefits of conserving their natural-resource base – much more so than large developers who don't live in the area' (MacKinnon 1995: 47).

While the interests of the community from a social standpoint are certainly paramount in the development of small-scale ecotourism, Williams (1992) feels that all factions within the community need to cooperate effectively in ensuring that a high-quality product is delivered without diminishing the ecology of the resource base. He advocates the development of an institutional structures strategy with the capacity to respond to tourism development before such development runs out of control. This strategy includes the following:

1 development of a grassroots planning process, driven by local interests and including aboriginal involvement;
2 understanding and appreciation of ecotourism market requirements;

3   an inventory of the region's resources to determine areas that are suitable for ecotourism and ones that are not;
4   the establishment of goals and objectives in line with concerns related to the cultural and natural impacts of ecotourism, with the creation of a vision statement to act as a control mechanism for the future; and
5   the establishment of a formal Tourism Management Board, to work with both the operators and the public, with the responsibility of monitoring change, communication, local benefits, etc.

Despite the fact that there will be direct and indirect beneficiaries and participants in the ecotourism industry, it is important to recognise that the entire community should benefit to some level. Participation therefore plays a key role in the initiative, as it empowers people to play a role in the decision-making process, where in many cases it is only those who are politically connected or affluent who are involved in the control and management of the enterprise (Sproule 1996). Such involvement can take the form of community representatives who speak for various elements of the population. This 'management committee' is very much like the proposed Tourism Management Board outlined by Williams (1992). The community ecotourism initiative needs to be founded on the notion of trust and transparency. Actions and decisions need to be communicated to the community through bulletin boards or other means. Money and finances are identified by Sproule as critical elements in dividing a community. To be open about expenditures and to share information in an attempt to dispel feelings of mismanagement or corruption is to be transparent in one's approach to management. Specifically, Sproule's community-based ecotourism (CBE) refers to

> ecotourism enterprises that are owned and managed by the community. Furthermore, community-based ecotourism implies that a community is taking care of their natural resources in order to gain income through operating a tourism enterprise and using that income to better their lives. It involves conservation, business enterprise and community development.
>
> (Sproule 1996: 3)

A final example of how management may be used to control the development of ecotourism initiatives within a community is outlined by Drake (1991). She defines local participation as 'the ability of local communities to influence the outcome of development projects such as ecotourism that have an impact on them' (ibid.: 132). Important in this process is the demonstrated benefit to the community both through community members' participation and through their realisation that some aspect of their community has been treated or protected (e.g. natural resources). The following is proposed as a model of local participation in the development of ecotourism projects (ibid.: 149–155):

- *Phase 1: Determine the role of local participation in project.* This includes an assessment of how local people can help achieve set goals through efficiency, increasing project effectiveness, building beneficiary capacity, and sharing project costs.
- *Phase 2: Choose research team.* The team should include a broad multidisciplinary approach and include people in the social sciences and those within the media.
- *Phase 3: Conduct preliminary studies.* Political, economic and social conditions of the community should be studied in the context of the environment, from existing documents and by other, survey-related, work. Identification and assessment of the following is important: needs, key local leaders, media, the community's commitment to the project, intersectoral involvement, traditional uses of the land, the type of people

interested in the project and why, the role of women, who will manage and finance the project, land ownership and cultural values.

- *Phase 4: Determine the level of local involvement.* Local involvement occurs along a continuum from low-intensity to high-intensity involvement. This must be determined in addition to when the involvement is to occur. In cases where government is not supportive of local government, intermediaries (e.g. NGOs) can be used to facilitate local participation.

- *Phase 5: Determine an appropriate participation mechanism.* This is affected by the level of intensity of the participation, the nature of existing institutions (e.g. government, NGOs, citizens' groups), and the characteristics of the local people (how vociferous they may or may not be). This phase may include information sharing and consultation, which usually takes the form of a citizen advisory committee with representatives from many groups within the community. The committee is charged with the task of commenting on goals and objectives or other project-related aspects.

- *Phase 6: Initiating dialogue and educational efforts.* The use of the press is important in this phase as a means by which to build consensus through public awareness. Key community representatives can be used in this process. The ecotourism team should explain the goals and objectives of the project, how the project will affect the community, the values of the area, any history of threats, and the benefits of the project. Various audiovisual techniques should be used to emphasise these points. Workshops or public meetings could be organised to identify strengths and weaknesses of the project.

- *Phase 7: Collective decision-making.* This is a critical stage that synthesises all research and information from the local population. The ecotourism project team present the findings of their research to the community, together with an action plan. Community members are asked to react to the plan, with the possible end result being a forum through which the team and local people negotiate to reach a final consensus based on the impacts of the project.

- *Phase 8: Development of an action plan and implementation scheme.* In this phase, the team and community develop an action plan for implementing solutions to identified problems. For example, if members of the community express the need to increase the community's standard of living, the team may respond by purchasing agricultural produce from local people at market rates or on a contractual basis. They may also develop a variety of positions to be occupied by local people, including gift shops, research positions, park management positions and private outfitting companies. This local action plan must then be integrated into the broader master plan of the project.

- *Phase 9: Monitoring and evaluation.* Monitoring and evaluation, although often neglected, should occur frequently and over the long term. The key to evaluation is to discover whether goals and objectives set out early in the project's life cycle have been accomplished or not.

Community development initiatives have a better chance of being accepted by local people if developers begin to acknowledge the fact that different groups within the community want different things, depending on their role in, affinity within, and utilisation of the community. This perspective is discussed by Jurowski (1996), who feels that because the impacts of tourism are not the same for all residents, residents' individual values need to be recognised by tourism developers in order for their projects to be successful.

The first unique group identified by Jurowski is the 'attached resident'. Such a person is likely to be a long-term resident or an older individual who loves living in the community because of its social and physical benefits. To these people, control over the form and function of their community is important. In general, tourism developers can gain support for their projects from this group by involving citizens in the planning process, establishing a focal point and common theme, developing projects that emphasise heritage themes, and showing that the project has social and ecological benefits for the community.

The second type established by Jurowski, the 'resource user', typically includes people like anglers and other recreationists who, although ambivalent about the economic impacts of tourism, can be won over by developers. The developers can gain their support by providing skill opportunities for youth, involving this group in events related to their interests (e.g. bike races), protecting 'their' sites for participation, and allocating tourism funds for the development of facilities and services they desire.

The final group outlined by Jurowski is the 'environmentalist'. Although this group is the most likely to focus on the negative aspects of the development, it is in the best interests of development teams to do the following: (1) provide information on how the project will protect the environment; (2) incorporate ecological education programmes; (3) encourage the participation of environmentalists in development; and (4) prompt citizens to develop their own educational programmes for tourists. Projects that reflect the interests and concerns of the community, therefore, are said generally to stimulate volunteer activity and minimise conflict.

This research is attractive because it acknowledges that people residing in the local community are indeed different from one another. This is in contrast to much tourism research which considers the local community as one homogeneous group (see also the work of Weaver and Wishard-Lambert 1996).

A key principle underlying the process of community development is the element of leadership. Mabey (1994) suggests that unlike the older model of leadership, seen as an individual influencing a group towards an end, newer leadership paradigms involve collaboration and partnerships between community individuals, groups, and organisations. This means devolving power to the followers, according to Mabey, and being fluid in the connection of people and resources. Belasco and Stayer (1993) likened this new style of leadership to what happens in a flight of geese:

> I could see the geese flying in their 'V' formation, the leadership changing frequently, with different geese taking the lead. I saw every goose being responsible for getting itself wherever the gaggle was going, changing roles whenever necessary, alternating as a leader, a follower or a scout. . . . I could see each goose being a leader.
>
> (Belasco and Stayer 1993: 18)

This example illustrates the importance of a common purpose in removing obstacles and establishing responsibility within the community ecotourism initiative. In essence the process of empowerment – holding the will, resources, and opportunity to make decisions within the community – allows people, either internal or external to the community, to step in and provide assistance in a respectful manner, while allowing local people to shape and control the pace of tourism development within their communities. Where this is especially relevant is in marginal places where there exists a tremendous void between those who can and those who cannot access information and resources. As such, education must play a key role in this empowerment process of revitalisation through ecotourism (e.g. providing the needed means to enable people to be informed of certain choices). An excellent example of education's role in empowerment lies in the Community Baboon Sanctuary project of Belize. Through education and partnerships, local farmers have stopped their conventional practice of clearing and denuding the land for agriculture purposes in favour of a land management strategy that is more environmentally sound. In so doing they have been able to attract international tourists fascinated by the ecological and cultural diversity of this part of the world (Lipske 1992). Education, therefore, helps to rekindle people's love and appreciation of the land, which is indeed important for both community-building and nation-building. A similar case study involves the Havasupai reservation adjacent to the Grand Canyon National Park in northern Arizona, where education and tourism have played a significant role in diversifying the local economy (White 1993).

The recent multidisciplinary focus on the concept of social capital may also provide the means by which to better understand roles and interactions in communities. Social capital may be defined as the ability of people to work together for common purposes within groups and organisations (Coleman 1988). Inherent in this general description is the notion that for people to come together over a common purpose there must be a great deal of social organisation, shared norms and values, all for the purpose of mutual benefit. For tourism, social capital may exist at a micro level as regards the social networks which exist within very tight social arrangements (e.g., people who must work closely together on a day-to-day basis), on a meso level with people throughout the community, and a macro level from an inter-community perspective (broadly speaking). Social capital is thus realised when these groups are able to come together in meaningful relationships which are predicated on trust, which in turn creates social cooperation, opportunity and stability. As suggested by Wyatt (1997), community is (1) an achievement, not a given, and (2) a function of what its members intend it to be. Ecotourism may be more effectively operationalised through the shared information, knowledge and interconnectedness between different stakeholders responsible for tourism development. While government agencies are often criticised for their silo, or independent, approach to business, a concerted effort to build bridges between agencies and between sectors within the community creates a stronger foundation for ecotourism development, and the impetus to employ successful models in other community initiatives.

## Partnerships

A partnership, as defined by Uhlik (1995: 14) is 'an on-going arrangement between two or more parties, based upon satisfying specifically identified, mutual needs. Such partnerships are characterized by durability over time, inclusiveness, cooperation, and flexibility.' More specifically, Uhlik (1995) developed a six-stage model of partnership development that concentrates on the conditions that will lead to a successful partnership agreement. These include (1) education of self and others; (2) needs assessment and resource inventory; (3) identifying prospective partners and investigating their needs and inventories; (4) comparing and contrasting needs and resources; (5) developing a partnership proposal; and (6) proposing a partnership.

Clements *et al.* (1993) suggest that because tourism is starting to be recognised as a community development tool, development must be sensitive to the requirements of many stakeholder groups, including tourism providers (e.g. hotels), public providers (e.g. recreation and park providers) and residents. Their view is that partnerships must be struck to ensure that a high-quality product is delivered, and is based on the notion that tourism experiences rely on all aspects of the community. The overall effectiveness of the delivery system is only as strong as its weakest link, and communities intent on the development of a tourism industry will increasingly rely on the positive benefits of partnerships in being accountable to the local and non-local public.

The partnerships developed for ecotourism, according to Sproule (1996), must also fit into systems that have been developed at regional and national levels. There are potentially many partnerships that can be struck to facilitate an atmosphere of cooperation and trust. Potential partners include: (1) organisations within the established tourism industry, particularly tour operators; (2) the government tourism bureau and natural resource agencies, especially the park service; (3) non-governmental organisations (NGOs), especially those involved with environmental issues, small business management and traditional community development; (4) universities and other research organisations; (5) other communities, including those with a history of tourism and also those that are just beginning; and (6) other international organisations, public and private funding institutions, national cultural committees, and many others; the tourism industry literature is replete with examples of stakeholder groups that have conventionally been at odds. In Canada, the United States, Australia, and many other countries around the world, there exists a significant degree of incompatibility between parks and aboriginal peoples. McNeely (1993) feels that despite what has occurred in the past, partnerships must be forged between these two groups in order that both may prosper in the late twentieth century and beyond. He feels that the following ten principles must be followed in order to help the cooperative efforts of these stakeholder groups (ibid.: 253):

1  Build on the foundations of local culture.
2  Give responsibility to local people.
3  Consider returning ownership of at least some protected areas to indigenous people.
4  Hire local people.
5  Link government development programmes with protected areas.
6  Give priority to small-scale local development.
7  Involve local people in preparing management plans.
8  Have the courage to enforce restrictions.
9  Build conservation into the evolving new national cultures.
10  Support diversity as a value.

Rather than emphasise what has not worked in the past, the partnership ideal must embrace the present and future needs of the groups involved in any transaction. For those in the realm of parks it may be biodiversity and the establishment of more protected areas. From the perspective of aboriginal people, decision-making and culture may be issues that top the list.

## CASE STUDY 8.1

### *Gorilla tourism in Africa*

The brutal 1994 campaign of genocide between the Tutsis and Hutus of Rwanda is a graphic reminder that conservation is very difficult to achieve without an understanding of the socio-economic context in which it takes place. Cultural differences prevent us from truly understanding another culture: their motives, lifestyles and moral bases. Stanford (1999) writes that the people of Central Africa have long suffered from too many governments, geographical divisions, colonial powers and dictators, and physical displacement. This situation, he suggests, has set the stage for an inherently flammable region, with Western aid and tourists adding fuel to the blaze. The 1998 capture of eight tourists in the Congo (four never heard from again), and the 1999 kidnapping of 14 tourists at the ecotourism centre in Buhomoa, Uganda (eight tourists murdered), brings to light the political and economic realities of this region. The importance of gorillas to the economy is clear. Despite the millions of people who were killed or displaced over the period of genocide in Rwanda, only five gorillas were killed (Stanford 1999).

*The Economist* reports in May 2002 that for the first time since 1985 poachers are trafficking in Rwanda's mountain gorillas, of which only about 600 remain. They note that two females were killed in order to capture their infants. An infant gorilla can fetch up to

*continued*

US$125,000 on the black market. The report acknowledges that the long-term benefits of gorilla tourism have always been substantial, with this form of tourism ranking as Rwanda's third foreign currency earner. The political instability put an end to gorilla tourism in Rwanda and the Congo; in Uganda, it has begun to rebound only after some significant initiatives by the Ugandan government to control poaching, preserve habitat, give 20 per cent of tracking permits to adjacent farmers, and use other funds to support the conservation budget of the Ugandan Wildlife Authority. In all cases, tour operators have invested substantial sums of money in the region, but this will not be a profitable area unless the government actively enforces policies which instil confidence in tourists and operators alike.

The International Gorilla Conservation Programme (IGCP) was developed with the goal to ensure the survival of the mountain gorillas and their habitat. The mountain gorilla is found only in protected afro-montane forests in north-west Rwanda, south-west Uganda, and eastern Democratic Republic of Congo. In these regions there are two populations. One is in the Bwindi Impenetrable National Park in Uganda, and the other in the ecologically homogeneous region of three parks (Mgahinga Gorilla National Park in Uganda, Volcano National Park in Rwanda, and Virunga National Park in the Congo). The programme is an initiative between three organisations, including the African Wildlife Foundation, the World Wildlife Federation, and the various protected areas authorities of the three respective countries. IGCP helps to monitor and protect the gorillas, train staff, and advise governments on policy and enforcement. They also help with habitat protection which, after poaching, is one of the biggest threats to the mountain gorilla in an area with tremendous population growth.

Gorilla tours cost about US$280 for the opportunity to visit a gorilla family for a period of 1 hour. As suggested below, tourists who have a cold or other illness are not allowed to make the journey. Primatologists argue that the health of the mountain gorillas is in danger because of the volume of tourists which they are exposed to. They note that tourists can expose the animals to human diseases, which may very quickly spread throughout the family (studies of gorilla faeces show that they have indeed picked up new parasites since the introduction of tourism), but it is not known to what extent wild populations are susceptible to diseases which have infected captive groups. The following from Tourism and Travel in Uganda (www.visituganda.com/wildlife/gorilla-rules.htm), are the rules and procedures that tourists must follow in their interactions with the gorillas:

- A maximum number of six visitors may visit a group of habituated gorillas per day.
- Voices are to be kept low.
- Do not leave rubbish in the park.
- Guides will inform tourists when to take pictures. Flash photography is not permitted.
- Always wash your hands before venturing out to the gorillas.
- Keep a minimum of 5 metres from the gorillas, to avoid exposure to diseases.
- Tourists must stay in a tight group whilst viewing gorillas.
- Do not smoke, eat or drink while near the gorillas.
- If a gorilla charges, crouch down slowly and do not look the gorilla in the eyes. Running will increase the risk of harm.
- Never attempt to touch a gorilla.
- If the gorillas become agitated, the one-hour tour may be cut short.
- After a visit with the gorillas, keep voices down until 200 metres from the family of gorillas.

Unfortunately Butynski (1998) observes that many of these directives are often overlooked in the process of conducting a gorilla ecotour. He reports that:

- The official number of tourists visiting gorillas has escalated from 6, to 8, with 10 being strongly considered, despite the cautions by scientific experts.
- All gorilla programmes (tours) suffer from a lack of risk assessments, impact assessments and programme evaluations.
- Tourists and guides have been found frequently to bribe park staff to ignore the rules. The benefit to the guide is a larger gratuity at the end of the tour, and likely a kick-back to park staff.
- Extended visits with gorillas have gone well beyond the one-hour limit.
- Sick tourists are included on tours and unauthorised visits to non-tourist gorilla groups have been allowed.
- Conservation has taken a back seat to political power-struggles and short-term financial gains.

Further, he notes that the absence of sound empirical data on gorilla tourism is the partial cause of so many ethical transgressions, and that information on the extent to which tourists effect gorilla behaviour, ecology, health and survival is essential. The sustainability of gorilla tourism is questioned because there is too much disparity between what needs to be done, and what the most important stakeholder groups (governments and NGOs) are willing or able to accomplish. This means putting conservation first over economic prosperity; conducting research; stronger regulations, which are enforced; and more money to support conservation. While tourism can be directly affected by war, conservation can be aided by it. A common conservation objective between nations – such as gorilla conservation – often forces these disparate parties to work together. Citing the director of the IGCP, Annette Lanjouw, Snell (2001) writes that peace is not a prerequisite for conservation. Lanjouw suggests that even though she works in a war-zone, conservation can contribute to regional stability.

Websites:

http://www.fauna-flora.org/around_the_world/africa/gorilla.htm

http://abcnews.go.com/sections/science/DailyNews/gorillas990305.html

## Aboriginal interests

As outlined above, the relationship between parks and aboriginal people has often been one based on conflict, and stems from the physical displacement and socio-economic fragmentation of aboriginal people. In Canada the relationship has been one of mistrust, where peripheral lands (occupied by aboriginals) have been run outside the control and sphere of influence of local native authority. This in itself is a perpetuation of the rift that has existed between Ottawa in the south and those living in the 'north'. Part of the problem exists because the concept of 'park' never existed in aboriginal languages, at least in North America, and the fundamentals of our Western view of park management (i.e. balancing recreational use with preservation) are completely alien to aboriginal culture, as outlined by Notzke:

> The distinction between work and leisure is largely an artifact of industrial society. The fragmentation and compartmentalization of life and environment which is part and parcel of western culture, is totally alien to aboriginal cosmologies, which view the world in an essentially holistic and unified manner.

(Notzke 1994: 1)

This view is supported by Sadler (1989), who suggests that aboriginals think of space primarily from a holistic perspective, which is contrary to the dichotomous Western perspective associated with conservation. Park management therefore proceeded to exclude aboriginals entirely from the park planning process. This has led to an indifferent, even hostile, attitude towards the existence of parks close to or adjacent to aboriginal lands. The Clearwater River Provincial Park in north-western Saskatchewan is an example of this indifference. Even though there is an abundant population of indigenous people in the vicinity of the park, there is little use of park resources (e.g. fish, game) by such people, in large part because of the park's political status (B. Wilson, personal communication, 4 December 1996). In order for parks to become accepted by the aboriginal population, therefore, some feel that a revision must be made to the preservation mandate that currently exists in some types of parks, changing it to one that is conservation oriented, which will be more consistent with the social, economic and ecological requirements of the First Nations people in Canada.

## CASE STUDY 8.2

### The struggle for Kakadu

According to Fox (1996), territories rich in resources provide more options for their use. This is the case with Kakadu National Park in the Northern Territory of Australia, which is a World Heritage Site and a place of immense ecological and cultural significance. Kakadu is owned by the Aborigines of the region, who lease it to the National Park Service. Legally, mismanagement of the resource may lead to the tenure being terminated. Ecotourism is thriving in the park, so much so that according to the Aborigines, there is no longer a distinction being made between the effects of mining and tourism; both are contributing to the degradation of the park. Some Aboriginal people, according to Fox, recognise that while mining will ultimately end, tourism will continue to grow and grow, and has manifested itself in the form of planned multi-million dollar tourism projects for the park. The issue of most concern to Aborigines is that their fundamental values are not consistent with current development philosophies in the park. According to Elliot (1991), mining and tourism in Kakadu is a perfect example of the competing and overlapping environmental values that various stakeholders have. For example, would it matter if our actions caused a species to become extinct? Would it matter if our actions caused the death of individual animals? Would it matter if we caused widespread erosion in the park? Is the extinction of a species an acceptable price to pay for increased employment opportunities? One would imagine that miners, the tourism industry, Aborigines, the park service, and other interests would all differ along these fundamental lines.

The element of control is critical in analysing any relationship between the ecotourism industry and indigenous people. For tourism development to be successful, aboriginal people must be allowed to take control of it despite the social and political forces that often work against them. This has been a problem in the restructuring of South Africa, where there remains a legacy of poor education and lack of training for black people (Swart and Saayman 1997). The result is that the whole touristic experience is one that is inherently foreign to the poor in South African society. Given that part of the government's tourism vision relates to addressing the need to improve the quality of life of every South African, Swart and Saayman recognise the need for significant changes in legislation to benefit all people.

Keller (1987), as discussed earlier, recognised that the Northwest Territories (NWT) has struggled with the development of its tourism industry, largely owing to the real lack of control of the industry at the local and regional levels. Keller's conclusions imply that the people of this peripheral region must have decision-making control and act to limit development to a scale of growth in tune with the social, ecological, and economic climate of the region. However, at least in some places in the NWT, Inuit, Indians and Metis people have started to realise that their varying social, ecological and economic needs can best be met through cooperative action with other agencies. The vehicle that is enabling them to do this is national parks, territorial parks, heritage rivers, and the tourism opportunities that emerge from such areas. For example, Seale (1992) suggests that two rivers, the Thelon and the Kazan, were easily designated as within the Canadian Heritage Rivers System, owing to the fact that the project planner lived within the community with his family for 18 months in an attempt to educate and work closely with local organisations such as the Hunters and Trappers Association and the Elders' Society. Seale believes that continued success will be contingent upon recognising the importance of conservation, tourism, and the needs of aboriginal people in Canada's north. In particular the following factors need to be considered:

## Community involvement

- Involvement of First Nations people in all aspects of tourism and conservation planning.

## Community benefits

- Continued and/or exclusive access to biophysical resources of the protected area for subsistence purposes.
- Provision of technical and professional training opportunities relating to positions in tourism and in conservation agencies.
- Priority status in hiring programmes undertaken by tourism interests and conservation agencies.
- Priority status in the licensing of businesses to be operated in the park or protected area.
- Compilation of traditional knowledge and heritage values of the aboriginal societies by the conservation jurisdiction, for use both by the communities themselves in strengthening their societal traditions, and by the conservation agency in managing the protected area and in giving to its visitors a heightened appreciation of the traditional society.

## Scale

- Ecotourism and adventure tourism can be accommodated much more readily than mass tourism, where control of such tourists (numbers and use patterns) will be in the communities' best interests socially and ecologically.

## Land ownership

- Tourism and protected areas will be accepted much more readily by aboriginal societies if the legal status of the land in question is first settled to their satisfaction.

## Sensitivity to needs of area residents and visitors

- Ecotourists must be informed as to what is and what is not acceptable behaviour in the north, in recognising that the quality of their experience depends in part on their compliance to accepted norms.

- Traditional societies and conservation organisations need to come together to establish patterns in which visitors and locals can meet in circumstances of mutual respect.

Increasingly, however, examples in the literature are emerging that demonstrate the positive signs of local indigenous control in the ecotourism industry. For example, Hitchcock (1993) illustrates that the semi-nomadic pastoralist people of Purros in Namibia have been able to develop a successful ecotourism industry. Ecotourists are asked to pay the equivalent of about US$12 to enter their tribal area, with all the money generated from tourism being divided equally among members of the community. Other aspects of the ecotourism initiative include (1) the fact that tourists must abide by structured guidelines (developed by the local people); (2) tour guiding and handicraft prices are set, so as to avoid undercutting; and (3) programmes to educate tourists on resource use in the area have been set up. Such community action, according to Hitchcock, is definitely on the rise in the developing world. Following point 1 above, Colvin (1994) proposes a code of ethics for the indigenous-sensitive ecotourist as set out in Table 8.1.

The issue of control in the development of ecotourism was at the centre of an incident between the Kuna Indians of Panama and the Panama national government. Chapin (1990) writes that in the 1970s the government of Panama had the specific aim of working with multinational corporations in the development of a tourism product in the region of the Kuna (around Carti on the Caribbean side of the country): 'In return for their services as tourist attractions, the Kuna were to be given employment as service staff' (Chapin 1990: 43). According to Chapin, the Kuna were unwilling to comply with the feasibility team and collectively forced the tourism industry influence out of the region. While plans for the development of casinos had to be discarded, one operator had his boat confiscated, and

**Table 8.1** *Code for the indigenous-sensitive ecotourist*

- Who operates the programme. Is it indigenous-run? If so, is it operated communally or do only a few individuals or families profit?
- If not indigenous-run, do local communities receive an equitable share of the profits or any other direct benefits, such as training? Or do only a few individuals/families benefit?
- Learn as much as you can about the local culture and customs. Visit local indigenous federation offices for information and materials with an indigenous perspective.
- Do not take photographs without asking permission.
- If you want to give a gift, make it a useful gift to the community rather than to an individual (e.g. gift to a school). Most indigenous communities function communally.
- Refrain from tipping individuals. If you are with a group, everyone can contribute to a gift to the community.
- Be aware of the boundaries of individual homes and gardens. Never enter or photograph without permission.
- Bring your own water purification tablets. Don't rely on boiling water exclusively as it depletes fuelwood and contributes to forest destruction.
- Pack out what you take in, and use biodegradable soaps.
- Be sensitive to those around you.
- Don't make promises you cannot or will not keep – for example, sending back photos to local people.
- Do not collect plants or plant products without permission.
- Wear appropriate and discreet clothing (e.g. many cultures are offended by women in shorts even though they may go topless).
- Respect local residents' privacy and customs. Treat people with the respect you would expect from visitors to your own home.

*Source*: Colvin (1994)

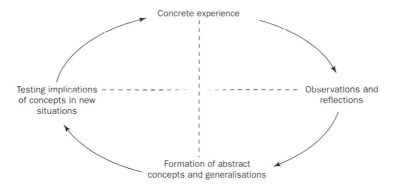

**Figure 8.1** Model of ecotourism educational training

*Source*: Wearing (1994)

another individual had to endure being shot, set on fire, a hanging attempt, and a pummelling at the hands of many Kuna youth. Conversely, the small hotels of the region that currently exist in the area are all run by Kuna and cater to a small and manageable number of tourists on the basis of Kuna ownership and Kuna law. Communities are therefore cognisant of the importance of sustainable development and are striving to achieve it, but under the following conditions: (1) they must have control of the industry; (2) projects must be small in size and therefore manageable; and (3) there must be equity and dialogue among the various stakeholders.

The importance of training and education appears to be a common thread that links the literature on ecotourism and indigenous development. Wearing (1994) emphasises that at any level of hierarchy, 'ordinary' people are perfectly able to solve their own problems if given the chance. Through an analysis of ecotourism and local indigenous communities, Wearing believes that experiential education and training are crucial in allowing such communities to develop ecotourism so that it is not just another commodified product where benefits are exported and natural resources are abused (Figure 8.1). 'Concrete experiences' are important in acting as cornerstones to teaching and learning in many indigenous communities. Aboriginal people can contrast this with the 'formation of abstract concepts and generalisations' attitude, which is more akin to the Western approach, and provides the means by which to bring a more conceptual foundation to the learning process. 'Observations and reflections' and 'testing implications of concepts in new situations', represent the integration of Western and local traditions that may be desired of an ecotourism experience, with the former representing a traditional method of learning in aboriginal cultures and the latter representing more of a Western scientific approach to problem-solving.

A significant body of research has surfaced in Australia on parks and Aboriginal involvement, where despite many years of agreements and talks there still exists uncertainty regarding the role that Aboriginals must play within protected areas. In Kakadu, Cobourg, and Uluru national parks, reported to be the model of Aboriginal–national parks relations in Australia, co-management strategies have not been successful in integrating and protecting indigenous values and rights (Cordell 1993), despite the fact that land in these areas has been given back to Aboriginals as freehold title. Davey (1993) reports that the models break down by failing to involve the Aboriginal landowners in policy and planning, by not having majority Aboriginal representation on park boards, and by the lack of training and skills to enable Aboriginal people better to understand their role within the park setting. Notwithstanding these failures, Australians have endeavoured to overcome some of these problems through research initiatives designed to develop Aboriginal skills and training. Baker *et al.* (1992) suggest that Aboriginal people can contribute equally to park management by applying Aboriginal ecological knowledge – which is based on generations of accumulated wisdom – with conventional biological surveys to overcome the time and finance constraints normally associated with research. In addressing the need to enhance Aboriginal skill levels (outlined above), De Lacy (1992) discusses the development of a number of Aboriginal employment programmes in Australia in the 1980s, including the initiation of many courses and skills that are currently available at selected universities within the Australian system.

Similar training opportunities have been reported by Liu (1994) in the American Affiliated Pacific Islands, where support of indigenous ecotourism entrepreneurs is thought to be critical in achieving sustainable economic growth. Liu suggests that government needs to take the lead in providing financing, development and operational support, and training programmes to develop appropriate marketing and business skills for this population. This is supported by King and Stewart (1996), who feel that national level controls are needed, in concert with local involvement, as a means by which to coordinate ecotourism development with the aim of reducing the effects of commodification of local indigenous cultures.

# Site development

The tourism industry has long been recognised as a formidable agent of social and ecological change (see Chapter 4 on impacts): some say in a positive context through increased education, renewed pride in culture, the conservation of heritage, and of course economics (Lickorish 1991); while others say in a negative sense, at the expense of people and the resource base. For example, hotel developments in Spain and Turkey have been identified as being excessive in their impacts on beach environments. In other cases, improper development and the lack of sewerage systems hamper the tourism industry to such an extent in places in Mexico and Brazil that tourists are told not to swim in the ocean owing to the pollution. In such cases tourism development is strongly affected by what does not exist (i.e. no sewerage systems) or because such systems are unable to keep pace with the ongoing increase in tourism visitation and associated development.

The style and extent of tourism development in the 1990s has been tempered by the trend towards the increase in the recent past in mega-development projects designed to cater to a growing market of travellers looking for self-contained, hassle-free vacations. Fennell (1989) writes that this mega-resort boom began in the mid-1980s, and includes sites such as the 31,000-acre resort community of Waikoloa in Hawaii and the 1,200-acre Sanctuary Cove mixed-use resort between Brisbane and the Gold Coast. Yet while most of the development has been centred in what have been politically stable democracies, management consultants have been looking to the untapped markets of the Far East (Fennell 1989; see also Ayala 1996).

In the midst of this mega-development push there has emerged the move towards more responsible travel and development, in response to the green movement (pressure from various groups including academics, ENGOs, and the general public), and in keeping one step ahead of the competition. The response to the changing emphasis has meant that even some of the major hotel chains have endeavoured to experiment with ecologically sound rooms through recycling, washing sheets only at the request of visitors, cutting down on the use of electricity, and so on (Andersen 1993; Wight 1995). However, where this changing focus has become the most noticeable is through the variety of small-scale initiatives that have developed in urban (bed-and-breakfasts), rural (farm tourism) and back-country (ecotourism developments) settings.

## Sustainable design and ecolodges

Over the past few years, several documents have emerged that analyse the development of ecolodges and ecotourism facilities on the basis of principles of sustainability. However, there appears to be some confusion on the scale and magnitude of development of such areas according to Epler Wood, who reports that ecotourism continues to be misunderstood in resort advertising schemes, while the term 'ecoresort' has yet to be appropriately defined (Shundich 1996). For instance, a tourism industry trade publication recently wrote that 'Another recent and perhaps surprising region for ecoresort development is India. . . . There is interest in genuine cultural experiences within the country. One man has proposed a historical tour of India by rail.'

Interest in sustainable ecotourism design took flight in the early 1990s in part as a result of a US national park service publication dedicated to the principles of sustainable design. Two events took place in the United States that prompted the development of this book. The first was the National Park Service Vail Symposium in October 1991 which discussed, among other things, the severity of environmental stresses within parks; the second was the Virgin Islands National Park initiative in November 1991, which specifically dealt with issues related to sustainable design in parks and protected areas (US Department of the Interior 1993). The latter publication explores interpretation, natural and cultural resources, site design, building design, energy management, water supply, waste prevention, and facility and maintenance operations in its overview of sustainable design. Of particular importance is its extensive checklist for sustainable building design that emphasises and respects the many interrelationships of all parts of the natural and cultural aspects of the site. However, these were not the first attempts to integrate the developments of humankind with the natural world (especially parks), as is evident in the seminal work of Miller (1976, 1989) on ecodevelopment.

Although the US Department of the Interior book is devoted to facility design and construction *in parks*, there is a more generic section that deals specifically with sustainable tourism development. Here the authors advocate using the principles of Aesculapia, the Greek place of healing. From this perspective, 'nature is respected for its restorative qualities. The human experience is set in harmony with the environment, and an opportunity is created to allow a reconnection of human needs to the natural systems on which all life is based' (US Department of the Interior 1993: 58). This type of design principle would then strive to meet the following criteria, according to the authors:

1    provide education for visitors on wildlife, native cultural resources, historic features, or natural features;
2    involve indigenous populations in operations and interpretation to foster local pride and visitor exposure to traditional values and techniques;
3    accomplish environmental restoration;
4    provide research and development on, and/or demonstration projects of, ways to minimise human impacts on the environment;
5    provide spiritual or emotional recuperation;
6    provide relaxation and recreation; and
7    educate visitors to realise that knowledge of our local and global environment is valuable and will empower their ability to make informed decisions.

Russell *et al.* (1995: x) define an ecolodge as 'a nature-dependent tourist lodge that meets the philosophy and principles of ecotourism'. Although they underscore the importance of the ecolodge concept from an educational and experiential perspective, they suggest that it is the philosophical tie with ecological sensitivity that must define these operations. The

**Table 8.2** *Traditional lodge versus ecolodge*

| Traditional | Ecolodge |
| --- | --- |
| Luxury | Comfortable basic needs |
| Generic style | Unique character style |
| Relaxation focus | Activity/educational focus |
| Activities are facility based (e.g. golf, tennis) | Activities are nature based (e.g. hiking, diving) |
| Enclave development | Integrated development with local environment |
| Group/consortium ownership common | Individual ownership common |
| Profit maximisation based on high guest capacity, services, prices | Profit maximisation based on strategic design, location, low capacity, services, price |
| High investment | Moderate/low investment |
| Key attractions are facility and surroundings | Key attractions are surroundings and facility |
| Gourmet meals, service, and presentation | Good/hearty meals and service-cultural influence |
| Market within chain | Market normally independent |
| Guides and nature interpreters non-existent or minor feature of operation | Guides and nature interpreters focus of operation |

*Source*: Russell *et al.* (1995)

importance of ecology is also underscored by the Ecotourism Society, which feels that such lodges promote an educational and participatory experience while at the same time being developed and managed in regard to the local environment in which it exists (Epler Wood in Shundich 1996). Russell *et al.* (1995) distinguish between traditional lodges and ecolodges using the 12 points shown in Table 8.2. They further distinguish between the ecolodge and the nature-based lodge, with the latter being associated with fishing, skiing, and luxury retreats.

The philosophical issue related to ecolodges identified above is one considered by Andersen (1993, 1994), who feels that along with environmental codes of ethics, a low-impact approach to the design of ecotourism facilities needs to be employed for such facilities to be truly sustainable – an approach that he suggests would entail a complete reworking of the conventional design of architects. Andersen's work is well publicised in the field of ecotourism, especially his advocacy of a number of principles which, broadly defined, include organisational issues (e.g. has an analysis been done of the area's ecological sensitivity?), site planning issues (e.g. minimise trail crossing points at rivers and streams), building design issues (e.g. maintenance of the ecosystem should take priority over view or dramatic design statements), energy resource and utility infrastructure issues (e.g. consider the use of passive or active solar or wind energy sources wherever practical), waste management issues (e.g. provide facilities for recycling), and evaluation (e.g. whether accommodation is made for older guests and physically disabled individuals). Andersen has been able to operationalise many of these design features in his Lapa Rios Resort development on the Osa Peninsula of Costa Rica's Pacific coast. This development has a main lodge, in addition to 14 private bungalows which were built by local engineers with the removal of only one tree. (See also Hadley and Crow 1995; Gurung 1995; Wight 1995 for good discussions on the guidelines and requirements for the design of ecotourism facilities.)

**Plate 8.1 (a)** There is no generic style for the construction of ecolodges. Main issues for consideration for ecolodges are the degree to which they are 'green' and the type of ownership (foreign and multinational, or local). **(b)** An example of a foreign multinational ecolodge, along with an environment which appears to be less ecologically authentic (manicured lawns etc.)

Given the low-capacity and low-impact philosophies of ecotourism development, a logical question is: at what scale does the ecolodge model *not* apply? That is, when is a development considered too big? This question, at least peripherally, has been addressed by Ayala (1996). As an international ecoresort developer, she believes both that ecotourism is becoming a segment of mass tourism and that it is not incompatible with beach tourism. She supports this view by suggesting that there is the need to diversify the vast number of sea, sun, and sand products through the addition of ecotourism (but not at the exclusion of other hard-path experiences). Ayala feels that the line between ecotourists and conventional tourists is least detectable in long-haul travel, citing similarities in their levels of education, household income and their occupations. Interesting is the fact that (1) the ability to travel long distances is directly related to one's wealth, and (2) wealth is directly related to education and occupation. Ayala fails to discuss other factors (e.g. psychographics) that determine differences in travel types. For example, the assumption being made is either that all long-haul tourists are ecotourists, or that all ecotourists are simply long-haul travellers. Ayala further writes that many of the mega-resort chains in Africa and Central America go so far as to donate money to conservation efforts. However, if the link is being made to ecotourism,

---

## CASE STUDY 8.3

### *Greylock Glen ecotourism resort*

This recently approved project is a consortium of state, local, not-for-profit agencies, and private interests which have come together to build a demonstration site for the concept of sustainable development in the Berkshire Mountains region of Massachusetts. The developers hope that this site will be a prototype for similar developments throughout the world. The resort will operationalise the concepts of sustainable development in offering educational, recreational, and conference-oriented programmes. In particular, visitors will be able to enjoy a golf course (developed from the Audubon Society's guidelines for environmentally sensitive golf courses), a nature centre (a year-round environmental education facility), hiking, Nordic skiing, biking, camping, tennis, swimming, horseback riding, and other recreational activities. In addition to a 200-room resort, 800 acres will be designated as open, recreational land. The developers (Greylock Management Associates) have specified that the following approaches to sustainability will be followed:

| | |
|---|---|
| Natural resources: | The maintenance of forest integrity and animal preservation will guide all development decisions at Greylock; a natural resource inventory will be compiled. |
| Cultural resources: | All cultural artefacts (e.g. farming fields, old roads, orchards) will be preserved and highlighted. |
| Energy management: | Energy meters will be present and interpret energy usage at the centre. |
| Water and sewerage: | Water-efficient toilets and shower heads will be used. Visitors will be presented with guidelines to help them cut down on water consumption. |
| Waste management: | A recycling programme will be implemented, in addition to composting. |

which is the focus of her article, then one wonders what percentage of the revenue from the hotels stays within the local communities and what percentage flows back to North America or Europe. Could such an initiative by the big companies be an appeasement mechanism or marketing scheme designed to look green? (See also the work of Thomlinson and Getz, 1996, on scale in ecotourism development.)

It is not easy to be sceptical about published work in this area, but increasingly scepticism is what is needed to address some of the issues related to the philosophy and implementation of ecotourism, especially as they relate to ecoresorts. Will ecotourism's fall from grace be directly attributable to the belief and practice that it fits very nicely as a form of mass tourism, and that mega-ecoresorts are the solution to bring ecotourism to the masses? Is this the rationale for their development – the intrinsic need to not deny the masses their place in the sun (or rather rainforest)? Or rather is it simply that from an extrinsic motivational perspective it is big business making money the old-fashioned way? Having seen some of these developments first hand, one merely has to dig a bit more closely at some of the unsustainable features of the development. For example, an ecolodge in Mexico recently developed by a large international company used grass and other materials not indigenous to the area in the design of the facility. In addition, no interpretive programmes had been developed to enable tourists to learn about or experience the natural and cultural attractions of the area. Such developments are merely cosmetic, often with few educational and sustainable features, which presumably are important to this segment, and undermine what Hawkins (1994) and others have suggested about ecotourism as being minimum density and low impact. Even in cases where ownership is in the hands of one or a handful of individuals, local ownership is rare.

## Ecolodge research

As outlined earlier in this chapter, ecotourism and the ecolodge concept have been misused by resort advertisers in the past. For this reason research needs to play a stronger role in providing leadership in this area. The following few examples serve to illustrate some of what exists in the area of ecolodge research.

Russell et al. (1995) undertook an international ecolodge survey of 28 operators in nine regions around the world (Belize, Costa Rica, Peru, Brazil, Ecuador, the state of Alaska, Australia, New Zealand and Africa). This study discovered that many of the lodges were found in or adjacent to protected areas, with outstanding natural beauty acting as a key to success of the operation. Most of the ecolodges sampled were small in size, accommodating about 24 guests, with some successful operations in Amazonia catering up to a hundred guests. Prices for accommodation (single night), meals, and local transportation were approximately US$130–170. The authors felt that although most of the ownership had been typically small scale and independent, corporate ownership was becoming more common. They cited the P&O line in Australia and the Hilton in Kenya as two examples of this recent phenomenon. Finally, they suggested that very few such lodges exist in North America owing to the existing reliance on camping and other, more comfortable recreational lodges. This finding is supported in a recent market study of ecotourism in Alberta and British Columbia (ARA Consultants 1994). However, according to Wight (1995), this is not necessarily what is happening. She feels that market research does not indicate that such a trend is due to market preference, but rather that it is due to the demand for a range of accommodation. To Wight, a better question would be to ascertain the range of accommodation types supplied by the region or destination area, where, she feels, there is an abundance of the conventional hotel/motel opportunities. She does, however, feel that there is a distinct lack of adventure-type, small-scale, fixed-roof accommodation (see Figure 8.2).

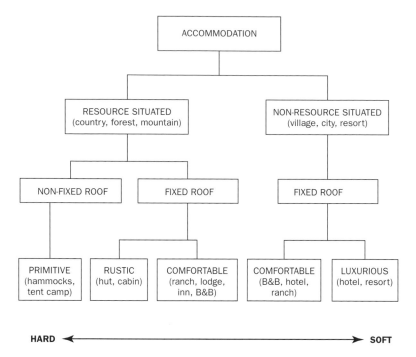

**Figure 8.2** Ecotourism accommodation spectrum

*Source*: Wight (1993a)

The hard aspects of ecotourism can be identified, at least in part, by non-permanent types of dwellings, including tents and hammocks. The ecotourism experience grows continually softer through fixed-roof units (cabins and lodges) on-site, and through fixed-roof dwellings off-site, usually in the form of hostels, hotels and resorts. This hard–soft accommodation continuum relates to the work of Laarman and Durst (1987), at least in part, who suggested that one of the hard and soft dimensions of ecotourism was the physical rigour of the trip (see Chapter 2). Those who were prepared to walk extensively in back-country regions, sleep in camps or crude shelters, or tolerate poorer sanitary conditions were said to have been interested in more of a hard-path ecotourism experience.

If this continuum is to apply to North America, it is logical to ask whether or not such operations are applicable to consumptive or non-consumptive forms of recreation. This issue has not been adequately addressed in the literature, at least from the Canadian perspective, and is an issue that is of the utmost importance as outfitters and operators wrestle with how to refit their enterprises to accommodate the changing recreational focus in the north, because, as Anderson/Fast (1996a: 54) suggest, 'depending on the characteristics of the package and the participating tourists, outfitter camps [the only form of accommodation in many remote regions of Canada's north] may or may not be appropriate'.

Finally, literature on ecoresorts seems to indicate that accommodation, although important, is not the overriding motivation for visiting regions. ARA (1994), for example, places accommodation below the natural setting and wildlife viewing (of primary importance), and other secondary activities related to the experience. To this, Wight (1995) adds support by suggesting that accommodation is an enabler of the overall ecotourism experience; basically tourists select the experience, then choose the accommodation. Often, though, prearranged package tours dictate much of the on-site experience, including choice of accommodation, restricting the opportunity for ecotourists to choose lodging, and reinforcing the point that there is an element of ambiguity as to just what ecotourism entails. According to Grenier *et al.* (1993: 9), 'one person's ecotourism dream may be another's touristic nightmare'.

## Conclusion

One of the unfortunate realities of ecotourism is that despite its altruistic intentions, to date there is little evidence that it is less intrusive than other types of tourism development. It is indeed frustrating that we continue to talk of appropriate means by which to control development, yet have very few successes to report on. In many localities around the world, local initiatives are chipping away at the conditions or circumstances that continue to plague ecotourism development. It is within the local arena that such change must occur.

There does seem to be pressure from international development firms to integrate the concept of ecotourism at a grander scale. While there certainly appears to be the opportunity to provide the tourist with a degree of the ecotourism experience at such a scale, caution needs to be used in promoting such areas as ecotourism sites. Where this relationship between mass tourism resorts and ecotourism breaks down seems to be in relation to the elements of ownership and control at the destination, with international or multinational domination of the industry still in existence. Regions intent on the development of eco-tourism must be able to secure both consensus within the community and the means, i.e. the decision-making responsibilities, to move away from many of the conventional global tourism development dysfunctions that are so prevalent within the industry.

## Summary questions

1   Explain some of the reasons why the benefits of tourism are far less to those service providers in the lesser developed countries than to those of the developed world
2   What is community development, and how is it that such a focus allows for a stronger element of control in the development of ecotourism?
3   Aboriginal interests have often been overlooked in development initiatives. Identify a few examples used in the chapter which help to explain how aboriginal people might benefit from ecotourism.
4   How do ecolodges differ from traditional tourism lodges (or hotels)?
5   Will ecotourism's fate be tied to megaresort and beach tourism, as some suggest? If so, how?

focusing on the subjective and personal dimensions of human life. Existentialists concern themselves with the question of the nature of being in understanding human existence. Using the example of geography, Tuan (1971) wrote that geographers establish contact with the world in two ways: nomothetically (environmentalism) and ideographically (existentialism). Environmentalism operates in a world of objects, existentialism in a world of purposeful beings. Whereas it is the goal of the environmentalist to search for general laws to discover meaning and order, the existentialist inherently strives for meaning in the landscape 'by tracing man-nature to its basis in man's bodily relation to the world' (ibid.: 185). (More on the ethical theories of deontology, teleology and existentialism, as they relate to tourism, follows.)

## Ethics and tourism

Researchers have for some time been instrumental in drawing attention to the various social, ecological and economic impacts of tourism in many regions of the world. Those regions experiencing the most pressure from tourism have been most vociferous in expressing their disdain of the tourism industry's presence. This was never more frankly implied than in the following statement by a Hawaiian delegate of a tourism conference: 'We don't want tourism. We don't want you. We don't want to be degraded as servants and dancers. That is cultural prostitution. I don't want to see a single one of you in Hawaii. There are no innocent tourists' (Pfafflin 1987: 578). Although this is an example of a social impact of tourism, more fundamentally it is a socially oriented ethical issue. We call attention to issues of this nature because we understand that there has been an injustice to either someone or a group of people. In the case of Hawaii, tourism is an economic cornerstone of the Hawaiian economy. To some people, its presence has contributed significantly to cultural dislocation within the region. Clearly, though, this is not the predominant sentiment that exists within Hawaii, especially from the perspective of those who stand to gain most from the industry. It is those who are realising the economic benefits of the industry who are in most cases supportive of its existence. For example, aboriginal people in Canada have started to come to realise the value of the tourism industry as a vehicle for economic development in their communities, where once they had been more sceptical. In many cases, their scepticism was founded upon the fact that they were not players within the industry, and were therefore not realising any of the potential economic benefits that accrue from the industry.

The presence or absence of acceptable ethical behaviour in tourism settings is very much a function of how tourists, operators, and local people act and feel about each other and towards the resource base. This can be seen through the following topology of interactions, as described by Fennell and Przeclawski (2003):

Tourist feelings towards:

- Self
- Residents
- Other tourists
- Brokers (those involved in various capacities within the tourism industry)
- The natural world

Resident feelings towards:

- Tourists
- Brokers

- Other residents of the community
- The natural world

Broker feelings towards:

- Tourists
- Residents
- Other brokers
- The natural world

There is the feeling among researchers that a balance must be struck between the various stakeholders of the tourism industry in ensuring that the goodwill of some stakeholders (e.g. tourists and local people) is not overridden by the misgivings of other stakeholder groups (e.g. government or industry). Obviously the sheer number of interactions between these different groups, as identified above, creates a myriad of cases where positive or negative impacts are generated. The three examples in Case Study 9.1 serve to illustrate this point.

## CASE STUDY 9.1

### Ethics in question: operators, local people, and tourists

On a recent trip to the jungles of the Yucatán Peninsula in Mexico, our group was taken to a farmer's residence before venturing off to see some of the famous Mayan ruins. At this farm the operator took great joy in showing us three jaguars that were secured to a fence using four-foot chains. The operator's rationale for showing us the jaguars was that he was certain that we would be unable to see jaguars in the forest over the course of our tour. He knew that such a sighting would be highly valued by ecotourists. This was his way of guaranteeing that we saw these magnificent beasts. Clearly, the animals did not mean anything to the operator, and he was taken back by the suggestion of one of our party that he would have rather seen a picture of a jaguar in a book than a live animal under these circumstances.

The second example involves the interference of local people in an ecotourist setting. While in Costa Rica on an ecotour some ten years ago, we had the pleasure, at a comfortable distance, of witnessing a green sea-turtle laying her eggs one night on a beach. After finishing her task she began to return to the sea, when suddenly a local man appeared and jumped on her back as she made her way. The utilitarian value that he placed on the turtle was clearly one that reflected the sentiment of Costa Rican people at the time, one that is not at all consistent with that of ecotourists. Turtles and turtle eggs are highly prized as a source of food by local people, even though many signs had been put up clearly stating that the personal consumption of turtle eggs was against the law (owing to declining stock).

The final example involves tourists visiting the great Ayers Rock of central Australia. A well-visited attraction, Ayers Rock attracts many different kinds of tourists over the course of the year. While upon Ayers Rock, again some ten years ago, I witnessed two tourists in their early twenties chip away pieces of this monolith as souvenirs to take home. It led me to wonder what the implications of this behaviour would be if all tourists had done the same. It would not take long for such effects to become highly noticeable.

In many cases it is the non-human members of our ecological community that are hardest hit by tourism (Figure 9.2). For example, in the game reserves in Africa a simple telephone call can organise 50 Land Rovers to converge on a pride of lions in a matter of minutes. It raises the question of equity in a holistic sense, which relates back to the discussion of human domination of the world set out early in this book. Should not all the planet's species have the opportunity to live in an untrammelled, unencumbered, natural state? This may very well be an impractical question, but one which needs to be addressed nevertheless (and is addressed in some of the more radical environmental philosophies such as Deep Ecology and ecofeminism). Myers (1980) illustrates that most people are concerned about issues like the disappearance of species, but not necessarily at the expense of other, more pressing matters such as population and pollution. Similarly, Elliot (1991) writes that there are degrees of environmental ethics to which people submit. These range in philosophy from human-centred ethics, which advocate a stance whereby ethics are evaluated on the basis of how they affect humans; to ecological holism, where the biosphere and the major ecosystems of the planet are morally considerable. Plants, animals, rocks, etc. matter only inasmuch as they maintain the significant whole. Other intermediary ethical stances include animal-centred

**Figure 9.2** Ethics and African game reserves

**Plate 9.1** This jaguar – chained up for 24 hours a day by a local farmer – was promoted as an eco-tourism attraction by the tour operator in Mexico

**Plate 9.2** In the turtle egg-laying season both tourists and local people must be sensitive to the needs of the female, which include space and the absence of harassment

ethics, where individual members of a species are the focus; life-centred ethics, where all living things are valued (but not necessarily equally); and ethics which imply that all living and non-living things have intrinsic value. Unfortunately, given the fact that tourists cross international, and hence cultural, borders, people do not always evolve an acceptable value-set in relation to the travel destination. This presents itself as a significant dilemma when one considers the adage: when in Rome, do as the Romans do – provided that what the Romans are doing is right (ethical relativism). One wonders whether there exists a more broad-based set of ethical values that ought to apply in all universal situations. This is a key issue that must be addressed in the tourism industry.

In the debate that has raged over the place of mankind on the planet, researchers have come to identify two main opposing paradigms. The first, anthropocentrism, is more of a dominant 'Western' world-view, while the second, biocentrism, is more of a minority, harmonious world-view (see Chapter 3 for a related discussion on conservation and environmentalism). The anthopocentric paradigm posits that nature can be conceived only from the perspective of human values. Humankind, therefore, determines the form and function of nature within human society. Conversely, the biocentric philosophy considers that all things in the biosphere have an equal right to exist, i.e. they are all equal in intrinsic value (Devall and Sessions 1985). In the realm of parks and protected areas, Swinnerton, in Philipson (1995), writes that resource managers and other decision-makers around the world have taken competing views on the place of environmental management in parks. Such views range from preservation to conservation to exploitation, and demonstrate the inconsistency in use of these terms (Figure 9.3). Those managers advocating the preservationist viewpoint adhere to more of a biocentric philosophy through the practice of little intervention, placing high value on natural resources, responsible use, and very small numbers of tourists.

There does appear to be a swaying of public sentiment in favour of a more environmentally oriented ethic according to Philipsen (1995). In citing the work of many authors (Dunlap

| | Resource protection | Resource development | |
|---|---|---|---|
| | Preservation ←→ | Conservation ←→ | Exploitation |
| View of resource | Biocentric/ anthropocentric | Anthropocentric | Anthropocentric |
| Level of intervention | No intervention | Limited intervention | Unlimited intervention |
| Measures of natural value | Undisturbedness Naturalness Completeness | Biodiversity Rareness | |
| Land use strategy | Segregation | Combination | Segregation/ combination |
| Access regulations | No use Responsible use | Controlled use Responsible use | Unlimited use Abusive use |
| | Very small numbers | Small numbers | Big numbers Mass tourism |

**Figure 9.3** Characteristics of resource protection and development

*Source*: Philipsen (1995)

and Heffernan 1975; Pinhey and Grimes 1979; van Liere and Noe 1981), Philipsen feels that the choices people make with respect to tourism reflect the importance of environmental values in their lives. Urry (1992), for example, writes that the collective tourist 'gaze' of the mass tourism phenomenon seems to be declining in the face of a more romantic gaze exemplified through green tourism.

## Too little research

Fundamentally, there is a very weak foundation of research in tourism ethics studies to date (D'Amore 1993; Payne and Dimanche 1996). There does however appear to be growing recognition within tourism's public, private and not-for-profit sectors that, from the perspective of sustainable development, ethical concerns have not fallen upon deaf ears. According to Hughes (1995), it is only the technical, rational, and scientific province of sustainability that has predominated to date, rather than the ethical province which was most responsible for the initial drive for a newer and more holistic development paradigm.

In the past, tourism ethics have been relegated to the area of hospitality management owing to the emphasis of hospitality's relationship to service and business (Wheeler 1994). This research provided the foundation for the move to establish the International Institute for Quality and Ethics in Service and Tourism (IIQUEST), which was designed to bridge the gap between ethics and issues related to community relations, sexual harassment, the rights of guests, and so on (see Hall 1993). However, the Rio Earth Summit of 1992 was a principal catalyst in generating more interest in the utility of ethics in the realm of tourism research – especially codes of ethics – where those in attendance committed themselves to Agenda 21. Genot (1995) outlines chapter 30 of this plan:

> Business and Industry, including transnational corporations, should be encouraged to adopt and report on the implementation of codes of conduct promoting best environmental practice, such as the International Chamber of Commerce's Business Charter on Sustainable Development and the chemical industry's responsible care initiative.
>
> (Genot 1995: 166)

Recently, a number of articles have surfaced either in response to the dearth of literature on tourism and ethics, and/or as a logical progression in the evolution of tourism research. This recent intensification is indeed timely as tourism researchers grapple with more philosophical issues that relate to business, society, and the environment. These additions to the literature include general discussions on ecotourism and ethics (Duenkel and Scott 1994; Kutay 1989), more specific commentary on ecotourism and ethics (Karwacki and Boyd 1995; Wight 1993a), tourism ethical decision-making on the quality of life in Third World countries, and ethics and marketing in the ecotourism industry (Wight 1993b).

A recurrent theme in many of the aforementioned publications is the advocacy of codes of conduct/ethics for the tourism industry. The British Columbia Ministry of Development, Industry, and Trade (1991: 21) defines a code of ethics as 'a set of guiding principles which govern the behaviour of the target group in pursuing their activity of interest'. From an industry perspective, they function as 'messages through which corporations hope to shape employee behaviour and effect change through explicit statements of desired behaviour' (Stevens 1994: 64). In tourism, such codes have gone beyond the realm of business to ensuring that local people, government, and tourists abide by predetermined guidelines. There has been a proliferation of such codes over the past few years by a variety of organisations, including government, NGOs, and industry, many of which can be found in the

**Figure 9.4** The Country Code

*Source*: Mason and Mowforth (1995)

work by Mason and Mowforth (1995) and the United Nations Environment Programme Industry and Environment (1995). An example of such a guideline is the Country Code, which was one of the first (Figure 9.4).

According to Scace *et al.* (1992) codes fall into two main categories: codes of ethics and codes of practice/conduct. The former, according to the authors, are philosophical and value-based, whereas the latter are more applicable and specific to actual practice in local situations. An example of a guideline falling within a code of ethics would be 'respect the frailty of the earth'; while an example of a code of practice guideline would probably be oriented more towards acceptable business practice with reference to the organisation's 'commitment' to the customer. This dichotomy, though, is somewhat misleading as it implies that codes of practice/conduct are not concerned with values. Although codes of conduct ought to be specific in actual situations, as suggested, they should also be founded upon a sound ethical principle as outlined in the previous discussion on ethics and business.

The link between the tourism industry and ethics has been discussed by Payne and Dimanche (1996: 997) who suggest that codes of ethics ought to project a number of key values, including justice, integrity, competence and utility, in illustrating that:

1   The tourism industry must recognise that its basis is a limited resource, the environment, and that sustainable economic development requires limits to growth.

2    The tourism industry must realise that it is community-based, and that greater consid-
     eration must be given to the socio-cultural costs of tourism development.
3    The tourism industry must also recognise that it is service oriented, and that it must
     treat employees as well as customers ethically.

Genot (1995) further underscores the practical reasons for industry to associate itself
with codes of conduct, stating that, among other factors, a sound environment means
good business, meeting consumer demand, unifying industry efforts and image, and
assuring product quality. Genot feels that the following principles are at the core of any
code of ethics: environmental commitment, responsibility, integrated planning, environmentally sound management, cooperation between decision-makers, and public
awareness. Many of these principles are elaborated upon in other publications devoted
specifically to the ethical conduct of operators and other members of the tourism industry
(see Dowling 1992; Ecotourism Society 1993). The Ecotourism Society publication, for
example, has developed guidelines for predeparture, guiding, monitoring, management and
local accommodations.

Stronger consideration needs to be given to the formation and understanding of these
codes, according to Malloy and Fennell (1998b). These authors subjected 40 codes (414
individual guidelines) to content analysis on the basis of who guidelines were developed for,
who guidelines were developed by, the type of tourism they were directed towards, the
orientation of the code, the mood of the message, and the main focus of the guideline. These
variables were juxtaposed with two philosophies of ethics (deontology and teleology), in
addition to determining whether the codes were relevant to a local ('local' included local,
regional, and national scales) or cosmopolitan (meaning a universal code) condition.
Deontology is an ethical approach which suggests that an act is right or wrong on the basis
of rules or principles of action or duties or rights or virtues (Mackie 1977). This approach
advocates behaviour that is means-based or intentions-based in its orientation. Deontology,
or right behaviour, provides us with guidance through the provision of rules and regulations
to follow; in essence, our 'duty' is laid out for us and we 'ought' to adhere to it. For example,
tour operators will abide by organisational policies developed for the purpose of honouring
established environmental or cultural norms. An example of a deontological code is as
follows: be aware of the periphery of a rookery or seal colony, and remain outside it. Follow
the instructions given by your leaders. Conversely, teleology, or good behaviour, is an ethical
approach which suggests that an act is right or wrong solely on the basis of the consequences
of its performance (Brody 1983). Because of its orientation towards the consequences of
one's action, it is ends-based. In this sense, the stakeholder is released from following the
tradition or dogma of the past and is able to choose in a manner which is consistent with
the changing circumstances of societies and cultures. An example of a teleological code is
as follows: Do not enter buildings at the research stations unless invited to do so. Remember
that scientific research is going on, and *any intrusion could affect the scientists' data*. (The
consequence of the action is highlighted in italics.) Figure 9.5 illustrates the methodology
employed in the study for each of the 414 guidelines.

In this research, approximately 77 per cent of all guidelines in the study were found to be
deontological in nature. As such, these guidelines, according to Malloy and Fennell, fail to
provide the decision-maker (e.g. tourist) with the rationale for abiding by a particular code,
with the assumption that an explanation of consequence (i.e. the teleological premise) is
unnecessary to the tourist or other stakeholder. More specifically, it was found that most of
the codes were developed by associations (NGOs), and for tourists; many of the codes were
ecologically based, rather than socially or economically based; about 85 per cent positively
stated; and the focus of such codes was on people or the resource base. Malloy and Fennell
recommend that future research in the area of codes of ethics should explore the deeper

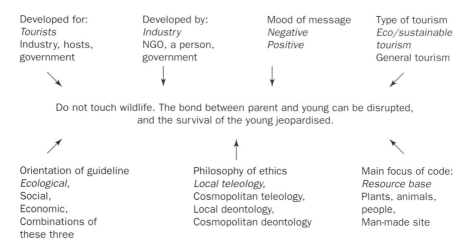

| Developed for: | Developed by: | Mood of message | Type of tourism |
|---|---|---|---|
| *Tourists* | *Industry* | *Negative* | *Eco/sustainable* |
| Industry, hosts, government | NGO, a person, government | *Positive* | *tourism* |
| | | | General tourism |

Do not touch wildlife. The bond between parent and young can be disrupted, and the survival of the young jeopardised.

| Orientation of guideline | Philosophy of ethics | Main focus of code: |
|---|---|---|
| *Ecological,* | *Local teleology,* | *Resource base* |
| Social, | Cosmopolitan teleology, | Plants, animals, |
| Economic, | Local deontology, | people, |
| Combinations of | Cosmopolitan deontology | Man-made site |
| these three | | |

**Figure 9.5** Example of an Antarctica code of ethics guideline

*Source*: Malloy and Fennell (1998b)

philosophical meaning of codes, advocate more of the teleological perspective in the development of codes (i.e. the consequences of one's action if the code is not adhered to), and should aim at a better understanding of the extent to which codes actually affect or change behaviour.

Despite what appears to be normative acceptance of the utility of codes of ethics in the tourism industry, it is clear that they have not been universally embraced by tourism researchers. Wheeller, for example, is not convinced of their utility, as demonstrated in a stinging review of ecotourism. He feels that there are no answers to the confusion surrounding both ecotourism and sustainable tourism, but simply

> a never-ending series of laughable codes of ethics: codes of ethics for travellers; codes of ethics for tourists, for government, and for tourism businesses. Codes for all – or, more likely, codeine for all. . . . But who really believes these codes are effective? I am pretty wary of platitudinous phrases like 'we are monitoring progress'. Has there been any progress – indeed, has there been any monitoring? Perhaps I am missing it and the answer itself is actually in code.
>
> (Wheeller 1994: 651)

## CASE STUDY 9.2

### *Ecotourism in Antarctica*

At 13.6 million square kilometres, Antarctica is the largest remaining wilderness on earth. It is covered by a sheet of ice which is on average 2.1 kilometres thick. While no humans live there permanently, the region is visited by hundreds of scientists each summer at one of 37 research stations operated by 18 different countries. Because of the unique location of Antarctica and its international significance, no one nation has sovereignty over any part of Antarctica. The region is governed by the Antarctic Treaty System of 1959 (came into effect in 1961), which provides for the conservation, research and management of Antarctic resources, and how these are to be used. As regards tourism, the Treaty in Recommendation VIII-9/1975 acknowledges that 'tourism is a natural development in this Area and that it requires regulation' (Heap 1990, cited in Bauer and Dowling 2003).

In a study of ship-based polar tourism, Grenier (1998) notes that the Argentinian ship, *Les Eclaireurs*, was the first to bring tourists to the Antarctic peninsula (the 1,600-kilometre spine that juts from the Antarctic continent towards South America) in 1957–58. In total, four cruises from Argentina and Chile brought about 500 tourists in that season. Lars-Eric Lindblad, however, is generally acknowledged to be the first to run international ecotours to Antarctica from New York (he chartered an Argentinian naval ship), starting in 1966. By the end of the 1990s, 84,173 ship-based passengers had visited Antarctica, with 14,623 alone visiting the region in 1999/2000, the most popular season ever (Bauer and Dowling 2003). The importance of ship-based tourism remains today, with up to 90 per cent of all tourists arriving this way.

Most trips to the Antarctic peninsula depart from Ushuaia, Argentina, Punta Arenas, Chile, or Port Stanley in the Falkland Islands, with the trip from the South American continent to Antarctica taking approximately two days. Ships accommodate anywhere from 35 to 400 passengers. Itineraries to this region of Antarctica vary according to the type of vessel used. Trips beyond the Antarctic peninsula, for example, require an ice-breaker. Those travelling to the Ross Sea section of Antarctica depart from Hobart, Tasmania, Invercargill, New Zealand, or Christchurch, New Zealand, from late December to late February, with mid-January to mid-February reported to be the best time to view wildlife. Ecotour activities on such trips normally include activities such as viewing penguins (some ships have helicopters with which to visit certain penguin colonies), whales, seals, and sea birds; visiting historic sites, scientific and whaling stations; and appreciating the grand scenery of the region. Tourists typically spend about two hours on shore at a time walking around and photographing wildlife.

Due to the absence of legislation regarding tourism in the Antarctic treaty, tourism operators, through the efforts of Lars-Eric Lindblad, took it upon themselves to develop codes of ethics for operators as well as for tourists. Lindblad also inspired a coalition of cruise operators, called the International Association of Antarctica Tour Operators (IAATO), to help administer these guidelines (Stonehouse 2001). The Rules for Antarctic Visitors, as distributed by Marine Expeditions, are as follows:

- Do not leave footprints in fragile mosses, lichens or grasses.
- Do not dump plastic or other non-biodegradable garbage overboard or on the continent.
- Do not violate the seals', penguins' or seabirds' personal space.
- Start with a baseline distance of 5 metres from penguins, seabirds and true seals, and 18 metres from fur seals.
- Give animals the right of way.
- Stay on the edge of, and don't walk through, animal groups.
- Back off if necessary.
- Never touch the animals.
- Do not interfere with protected areas or scientific research.
- Do not take souvenirs.

Despite these guidelines, Grenier (1998: 186) has put together a list of some ethical transgressions that he has observed in his polar travels over the course of many years. These include:

*continued*

- A musician scaring away a group of penguins by giving his 'first Antarctic flute concert'.
- Passengers collecting stones, feathers and bones for souvenirs.
- A passenger tossing stones at the foot of a penguin to improve a photographic opportunity.
- Visitors touching penguins.
- A tourist walking straight into a penguin rookery, despite warning, stating that he had 'paid for the right to go as he pleased'. The penguins were scared off, exposing chicks to predators and freezing temperatures.
- A captain positioning himself next to a seal for a better photo opportunity.
- A zodiac driver consuming alcohol before an excursion conducted after sunset.
- A crew member harassing juvenile penguins dying after having been abandoned by their parents.

Davis (1998) notes that the management of people in this type of wilderness is difficult because of the lack of sovereignty by any one country, and because of the differing cultural views on wilderness management philosophies. Tourists have consistently shown interest in the development of more facilities on the continent – like toilets, accommodation, post office, gift shop, and other activities such as diving, skiing and camping. Davis's preference is to stipulate which activities are acceptable as opposed to regulating all activities equally.

Hall and Wouters (1995) observe that because of the fact that only 2 per cent of the land is ice free, there is a tremendous amount of competition between humans and wildlife, the latter using this area for breeding. This is compounded by the fact that the tourism season coincides with the peak breeding season. The constant human intervention is thought to cause behavioural changes in animals, denudation of habitat, and the spread of new organisms which might have the potential to spread animal and plant diseases. In 1997, for example, a poultry virus, brought to the continent either by scientists or tourists, was found to have infected Antarctic penguins. The concern is not only what types of impacts are occurring on Antarctica, but who causes them. Tangley (1988) writes that while scientists worry about the impact of tourism, conservationists say it is science that threatens the fragile Antarctic environment. This issue has touched off a tremendous debate on the responsibilities and effects of both parties, with scientists most vociferous about the dangers of an ever-increasing presence of tourists on the continent.

Like the Galápagos, Antarctica is truly one of the world's prime ecotourism hotspots. How it comes to be managed is absolutely critical as a case study which might provide meaningful leadership in the management of other tourism regions around the world. In considering this region, the question of humankind's obsession with visiting every corner of the planet comes to mind. Some argue, as scientists in Antarctica might support, that tourists have no reason or purpose to be in Antarctica. Others suggest that it is our right as humans to go where there is opportunity to see and experience new lands. The question might become more important over time as we see tourism continually expand into the world's most pristine areas. The argument for tourism in these areas is the notion that it helps to support economically depressed communities. Antarctica is perhaps the only place in the world where this position does not hold true.

Websites:

http://sedac.ciesin.org/pidb/texts/antarctic.treaty.protocol.1991.html
http://astro.uchicago.edu/cara/vtour/
http://www.lonelyplanet.com/destinations/antarctica/antarctica/facts.htm

# Beyond codes of ethics

As stated earlier, the base of ethics and tourism/ecotourism research is at present quite limited. Given this fact, it is the purpose of this section to highlight some of the work in this area as a means by which to provide a basic understanding of the extent and appropriateness of this work to the future of ecotourism studies.

In a well-prepared essay on ethics and tourism, Hultsman (1995) proposes the use of a framework of ethics and tourism based on the notion of 'just' tourism, which refers not only to acting in a fair, honourable, and proper manner, but also to the fact that tourism is 'merely' or 'only' a 'small thing'. With reference to the latter aspect of the framework, Hultsman emphasises that tourism has become more important as an economy than as an experience, and as such has lost some of its intrinsic qualities that are derived through the pleasure of experience (see Chapter 7). He notes that 'Should tourism reach the point of being considered by service providers as first a business and second an experience, it is no longer "just tourism"; it is industry' (ibid.: 561). He concludes his paper by suggesting that ethical issues need to be included in the textbooks used in tourism curricula, and further that, like many other authors, we ought to be concerned that there is a real lack of professionalism in the delivery of tourism services.

In other related research, Upchurch and Ruhland (1995: 37) focused on advancing the hospitality industry's understanding of ethical work-climate types through an analysis of the normative ethical theories of Egoism ('an individual should follow the greatest good for oneself'), Benevolence ('actions that are delivered on a fair and impartial basis, are based on maximizing the good, and follow impartial distribution rules'), and Principle (which 'is a theory suggesting that decisions are based on rules. The outcome or actions should be based on the merit of the rule'). Respondents were Missouri lodging operators who filled out the Likert-style ethical climate questionnaire designed to measure the three ethical theories. These authors found that Benevolence was the most frequently perceived ethical

**Table 9.1** *Ethical orientations: a comparison*

| | *Social obligation no. 1 (Friedman)* | *Social responsibility no. 2 (Davis)* | *Social responsiveness (extension of No. 2)* |
|---|---|---|---|
| General overview | Legal and profitable | Current social problems are responded to | Future social and/or environmental problems are anticipated and addressed |
| Choosing options | The sole consideration aside from profit is legality | Decisions respond to social issues which overtly need to be addressed | Decisions based on anticipation of future needs and/or social problems even if they do not impact or are caused by the firm |
| Strategies are evaluated with reference to: | Is the strategy legal? Is the strategy profitable enough? | Has the organisation responded to problems and issues which have emerged as significant? | Future problems are addressed even if the organisation is not directly causing them |

*Source*: Walle (1995)

climate type present in the lodging respondents (which, they say, is consistent with the literature). Benevolence, therefore, indicated a certain responsiveness to the clientele in a socially oriented manner.

This element of social responsibility is also discussed in a paper on business ethics and tourism by Walle (1995). He discusses the dichotomy that exists in business ethics theory as established by the work of Friedman and Davis. He illustrates that for the Friedman school, the *modus operandi* of business is to generate profits, and that it is not the responsibility of business leaders to dwell upon social policies and strategies (within the law). On the other hand, he argues that the Davis school advocates a more socially responsible behaviour for business, which will in turn lead to greater profits and prevent government intervention, through the generation of positive publicity (Table 9.1). Through these extremes, Walle argues that tourism, because of its uniqueness, cannot follow the universal or generic strategies of mainstream business which focus on the organisation and its customers (e.g. manufacturing). Instead, he makes the point that

> Tourism is not a generic industry since it uniquely impacts on the environment, society and cultural systems in ways which require a holistic orientation within a broad and multidimensional context. Contemporary business ethics, however, has been slow to embrace such a holistic perspective. Historically, the focus has been on the organization and its customers. Impacts on third parties (externality issues) have often been ignored.
>
> (Walle 1995: 226)

**Table 9.2 *Special ethical considerations of tourism***

| Tourism's perspective | Social obligation no. 1 (Friedman) | Social responsibility no. 2 (Davis) | Social responsiveness (extension of no. 2) |
|---|---|---|---|
| Progress is not inevitable or inherently beneficial | Since the concept of 'progress' is not universal or inevitable, we should not place an over-reliance upon it in our strategies/tactics | Tourism has a responsibility to encourage development which meshes with the local environment and culture, not in accordance with a universal concept of 'progress' | Since 'progress' leads to concomitant changes in culture and the environment, tourism strategy should be appropriate and mitigate its impact |
| Tourism can be undermined by pressures of the industry | Change wrought by tourism might undercut the industry. Such potential should be prevented and mitigated when doing so is a good tactic | Tourism causes negative impacts and pressures on people and the environment which should be mitigated | The industry has both practical and ethical reasons to respond to impacts on the environment and local people |
| All relevant stakeholders need to be considered when strategies are forged | Government regulation and loan conditions might demand responding to the needs of all relevant stakeholders | Tourism should respond to the needs of various stakeholders which are impacted on by the industry | The industry should anticipate future impacts from various sources and respond in proactive ways |

*Source*: Walle (1995)

His initial conceptualisation on social obligation, responsibility, and responsiveness is reworked to illustrate how such orientations might be applied to tourism's uniqueness (Table 9.2).

In work related more specifically to ecotourism, Fennell and Malloy (1995) drew a relationship between meta-ethics (existentialism) and normative ethics (deontology and teleology) in discussing how each can contribute to comprehensive ethical decision-making in the ecotourism industry. This, they reason, is a deviation away from the traditional manner by which to resolve ethical decisions, i.e. mediating between the two dichotomous approaches of meta-ethics and normative ethics.

Existentialism implies that an act is right or wrong according to the actor's free will, responsibility, and authenticity (Guignon 1986). Existentialism or authentic behaviour favours neither ends nor means, but rather the premise that individuals need to be self-aware and prepared to take full responsibility for their actions. In essence, this provides the opportunity for individuals to choose freely their behavioural patterns within the confines of their own acceptance of responsibility for all consequences on all things. Deontology and teleology have been discussed above.

Fennell and Malloy go on to suggest that although each of the three (existentialism, deontology and teleology) advocates radically different perspectives, stakeholders or decision-makers in search of comprehensive ethical decisions may employ any of them to arrive at ethically good, right, and authentic solutions. Also, according to these authors, it is not the case that people rely exclusively on one pure form of ethics as a means by which to make decisions. They feel that what might be termed a triangulated approach would correspond to the many demands, both organisational and moral, that exist within the tourism industry (Figure 9.6). For the ecotourist, this would accomplish two things:

First the individual is provided with an ethical criteria [*sic*] with which to assess and resolve issues in a comprehensive fashion that is normative as well as introspective. As a result, it is hoped that what for many may be a latent sense of ethics, may become more apparent and considered as a result of exploring teleological, deontological, and

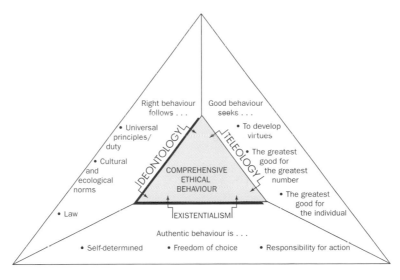

**Figure 9.6** A model of ethical triangulation

*Source*: Fennell and Malloy (1995)

existential perspectives. A second consequence of this model is that it is used as a point of departure for not only empirical research of the behaviour of ecotourists, but also for the development of more comprehensive ethical conduct for the ecotourism industry in general (e.g. organisational culture, climate).

(Fennell and Malloy 1995: 178)

In other research, Malloy and Fennell (1998a) examined the organisational culture literature in attempting to differentiate between ethical and non-ethical work climates in the ecotourism industry (Figure 9.7), on the basis of work by Schein (1985) and Kohlberg (1981, 1984). Schein wrote that culture sets certain standards of acceptable behaviour within an organisation. Individuals are thereby subject to a process of socialisation based on normative and value-based behaviours of the organisation. On the other hand, Kohlberg's work is fundamental to our understanding of moral development within society from the perspective of three distinct phases: pre-conventional (people who act to avoid punishment, receive rewards from external sources, seek pleasure, and give little consideration to social norms and ecological principles); conventional (individuals act to gain approval within society by adhering to social sanctions); and post-conventional (defined by the move to be influenced not by external forces but rather from within for the good of the community, but also, more broadly, for the good of humanity and the planet). Malloy and Fennell (1998a) termed these three phases, applied to ecotourism, respectively: (1) the market ecotourism culture, (2) the socio-bureaucratic ecotourism culture, and (3) principled ecotourism culture. According to these authors, it is this last stage that the ecotourism industry must strive to reach, which demands not solely an economic and/or sociological agenda, but rather a socio-ecological change that reflects the general goals of ecological and social holism. The transitions between stages are influenced by broad political processes (e.g. the development of policies to control tourism impacts); cultural processes (e.g. promotion within firms based

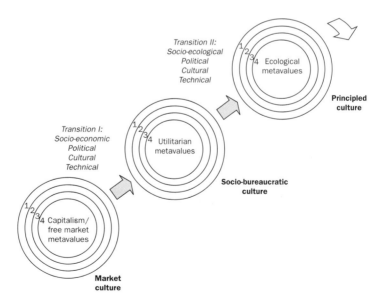

**Figure 9.7** Moral development in ecotourism organisational cultures. 1, Artefacts of the ecotourism organisation; 2, patterns of individual behaviour in the ecotourism organisation; 3, values and beliefs of the ecotourism organisation; 4, basic assumptions of the ecotourism organisation

*Source*: Malloy and Fennell (1998a)

on socially and ecologically responsible behaviour); and technical processes (e.g. certification of ecotourism guides).

## Conclusion

One of the concerns that many people in the field of ecotourism share, at least from the research standpoint, is the lack of empirical data to substantiate many of the claims being made about ecotourism as being more responsible than other forms of tourism. For example, is ecotourism a greener form of tourism than other forms? Are ecotourists more ethically based than other types of tourists? What about their values and attitudes? These and other questions must be addressed in order to profile this type of tourism more accurately. A potentially fruitful area in differentiating ecotourists from other types of tourists is through an analysis of values. Conventional measures used in other social science fields include Rokeach's Value Survey, the Values and Lifestyle Scale, and the List of Values Scale (see Madrigal and Kahle 1994; Kahle *et al.* 1986). Madrigal (1995) writes that owing to their centrality to a person's cognitive structure, personal values, and therefore scales of this nature, are effective predictors of human behaviour. In relation to attitudes, the New Environmental Paradigm (NEP) scale has been moderately successful at measuring beliefs about humankind's dominance and harmony with the natural world (Jurowski *et al.* 1995), and may also lend some support in differentiating between, for example, ecotourists and mass tourists within a destination. Finally, in the context of the business literature, Reidenbach and Robin (1988, 1990) have developed a Multidimensional Ethics Scale (MES) which is designed to determine the ethical differences between subjects on the basis of deontological, teleological, justice, and relativistic theories of ethics. Using this scale, Fennell and Malloy (1997) found that ecotour operators were moderately more ethical than other types of operators (fishing, cruise line, adventure and golf) on the basis of their responses to three tourism scenarios, providing some empirical evidence to suggest that, at least from the perspective of operators, ecotourism is in fact a more ethically based sector of the tourism industry.

## Summary questions

1   What is the difference between theoretical and applied ethics, and why are they so important together?
2   Speculate as to why there has been such a reluctance in taking a more proactive role in understanding how ethics might benefit the ecotourism industry.
3   How do deontological and teleological codes of ethics differ? Which is better, and why?
4   What is ethical triangulation?
5   Which code of ethics is reported to be one of the oldest, as regards tourism and outdoor recreation?

# ⬤10 Conclusion

We come more and more to see through the follies and vanities of the world and to appreciate the real values. We load ourselves up with so many false burdens, our complex civilization breeds in us so many false or artificial wants, that we become separated from the real sources of our strength and health as by a gulf.

(J. Burroughs, *Time and Change*, 1912)

This book has attempted to project a balanced approach to the foundations, concepts, and issues of ecotourism as they currently exist. At the outset, it was mentioned that there are grounds for maintaining either a positive or negative outlook on where ecotourism has been and where it is going. Perhaps this is the place for more of a personal, introspective look at the phenomenon.

There can be no question that disparate views exist on how and at what level ecotourism ought to be conceptualised and delivered. Like so many other things, it depends on the angle from which one approaches the subject (e.g. economics, marketing, ecology). Admittedly, the approach taken in this book is one that is more protectionist, which has ultimately led to a relatively elaborate definition of ecotourism. Such a view is based on the belief that if ecotourism is to survive it will have to do so by reverting back to or adopting a 'greener' agenda. While some argue, as I do, for a stricter definition of ecotourism, some of my colleagues go so far as to suggest that ecotourism cannot survive without a relationship with sustainable mass tourism. We will have to see. If ecotourism does partner mass tourism, has it not undergone a significant transformation? Is it not then mass sustainable tourism with a sprinkling of nature? I was told as a graduate student that if ecotourism were to become successful, it would have undermined everything that it had initially set out to accomplish. To me this is fast becoming reality. Those who see ecotourism as the way of the future for all tourism, however, must take heed of the words of Wall (1994), who wrote that:

> The most important environmental challenge facing tourist planners and managers is not to find a means of inserting small numbers of environmentally aware visitors into pristine environments (although this is certainly a worthy objective); rather it is to devise sustainable forms of mass tourism.
>
> (Wall 1994: 9)

This quotation emphasises the fact that sustainable tourism must lead the way in the future development and management of the tourism industry. Ecotourism is but one type of sustainable tourism (in principle), against a backdrop of many, many other forms. Wall further asks the question, which has been addressed earlier in the book, of whether forms of mass tourism can be developed which are environmentally benign. The research and the many case studies around the world illustrate that mass tourism will have to become more sustainable, especially in light of the growing magnitude of the industry – is there any end in sight? If ecotourism has a role to play in mass tourism's transformation it will be by demonstrating, at a micro scale, the ability of the industry to become more ecologically accountable and responsible through the development of alternative energy strategies, sensitivity to the resource base and local people, and so on. Crossing over the line, however, may have disastrous consequences. A majority role in this capacity to change will be played by those involved in sustainable tourism, e.g. Europe, where some of the best research and development in the industry is occurring. We cannot afford to lose sight of how the communities of Europe have initiated many superior programmes and policies that advocate doing more with less.

The issue of mass tourism as ecotourism pertains to a related problem surrounding ecolabelling and resort development. Personally, I worry when I see names like Ramada on ecolodge signs, which is the case in a recently developed ecoresort in the Yucatán Peninsula of Mexico. This concern is not necessarily from the perspective of ecological impact (Ramada has apparently adopted an environmental integrity image in its product development, as have other hotel giants such as Holiday Inn), but rather from a socio-economic perspective. One of the tenets of ecotourism was that it is supposed to be locally oriented. How will Ramada remain internationally competitive if it is pouring its revenues into the local community and the preservation of biodiversity of the region? Flares are automatically discharged when the international business community gets involved in such 'local' enterprises. This scenario is more along the lines of what Britton (1977) and others suggest about metropolitan domination of the industry. In my estimation, one of the biggest threats facing the ecotourism industry today, as implied earlier, is the danger that the large hotel and resort corporations will infiltrate the ecotourism industry with less than altruistic ends.

Canadian Pacific Hotels and Resorts, for example, has recently developed more environmentally friendly policies as a result of a formal survey of its thousands of employees (D'Amore 1993). Its 12-point environmental action plan is indeed a laudable example of leadership in this area. Would such policies have appeared popular 15 years ago? Not likely. Ramada and other hotel chains are responding to what they perceive as a movement in public opinion on the management of hotels. There appears to be a good deal of business savvy in this choice (i.e. they benefit the hotel in terms of lower maintenance and overhead costs, but also the customer in the sense of peace of mind), but is that all it is? It reminds me of the recent trend for big companies to provide health memberships for their workers on the assumption that there is a genuine interest in the health and well-being of employees. The underlying principle of this provision may be that employees are encouraged to get fit and in the process be more productive on the job with less absenteeism, and with the ultimate end of saving the corporation money. In cases such as this where there seems to be a win–win situation, it is hard to argue against the fact that the corporation is doing nothing more than treating the worker as a commodity with the ultimate goal of extracting the maximum yield out of this resource. I do not wish to be too pessimistic, but it is difficult to determine whether the motives of the hotels are truly intrinsic (principled actions) or extrinsic (economic gain) in nature. One wonders about the lengths to which hotels and other businesses would go to remain successful if resource protection and environmental concerns were not contingent upon financial success and public opinion.

A key issue to address in the future is the role that government and industry play in the delivery of the ecotourism product. There is no question that the private sector, when left to its own designs, will strive to make money. Given this axiom, if ecotourism fails, should the responsibility for its demise lie squarely on the shoulders of the operators? Not directly. More appropriately, I believe, fault should lie in the hands of national governments and international organisations involved in directing tourism development for their inability to provide leadership and implement appropriate guidelines or regulations within the industry.

In addition, government-sponsored research must be completed responsibly. In some cases, marketers and consultants have seen fit to rationalise ecotourism's market potential and therefore overall economic impact through irresponsible methodological approaches, which has, in part, contributed to the erosion of the concept. We have allowed marketing to dictate experiences rather than vice versa.

Some of the critics of ecotourism argue that (1) it must operate at low levels or carrying capacities in order for it to become successful, and as a result (2) it may not be sufficient to allow all members of a community to benefit economically. Communities are defined as such by the diversity of human and physical resources that they comprise. Just as in ecosystems, members of the community have roles to play and niches to occupy. The same principle holds true in an economic context. Development theory posits that economic viability is very much a function of a balanced approach to economic development. Just as monocultures are a threat in an ecological context, regions bent on establishing ecotourism as the only form of economic development are subject to the same dysfunctions. Such regions should be well aware of the fact that ecotourism must be integrated into a broader sustainable development policy, which constitutes strong planning and management of the resource and the industry. Both planning and management are critical to the delivery of ecotourism, and must occur through the institutional arrangements that have been put in place to allow for it to occur, i.e. rules, laws, regulations, and so on – which is leadership that must come from government. A poorly run industry does not a poorly conceived idea make.

Critics of ecotourism further charge that the definitions and goals of ecotourism are far too lofty, i.e. there is very little chance that all aspects of elaborate definitions can be reached. This argument is a valid one that nevertheless should be examined. Essentially there are two ways of dealing with it: (1) to agree, and modify the definition to be more forgiving, which is the prevailing norm; or (2) to argue that we must set lofty goals in order to accomplish prescribed ends. In support of the latter, the rationale is that the establishing of certain objectives, or principles of ecotourism, induces practitioners and researchers to be true to these ends. In addition, it enables, in much the same way as the design of the accreditation programme in Australia, the attainment of a means by which to separate operators on the basis of established criteria. Elitist? Not really; it provides the consumer with needed information in making informed decisions on product content.

Oscar Wilde said that a critic is 'a man who knows the price of everything, and the value of nothing'. In my estimation, the real job that lies ahead is not in writing off ecotourism (this is too easy), but rather in striving to build clarity and meaning into the concept through some fundamental restructuring. Perhaps we just need to play better by the rules, but in the case of ecotourism the rules have yet to be fully agreed upon. Sustainable development has its critics, but at the same time there are few other options to look to for a greener future. Will there be any new substitutes for ecotourism to look to on the horizon (e.g. 'deep ecotourism'; geotourism which at the time of press is making the rounds as another type of nature/cultural tourism phenomenon)? Most likely not. Ecotourism was supposed to be tourism's Deep Ecology. As yet, according to many case studies around the world, it is not, and it will not be until we address some fundamental issues related to its philosophy and delivery. The vision for the future of ecotourism (and sustainable tourism) is not universally shared either from the perspective of the industry (tourists, local people, and the industry) or geographically – as a collective calling or dream. We must view the environment and its relationship to tourism as being like a bank: we keep withdrawing without making any deposits, and while such deposits could be earning valuable interest in the form of a stronger human–land relationship and a better way to conduct commerce, the withdrawals we make serve to intensify the dichotomy that exists between people and the natural world. To this end I have heard ecotourism referred to as being rather like a game of pick-up sticks. One must carefully isolate a stick from a pile of sticks and remove it without disturbing the others. The analogy to be made is that when tourists venture off into wilderness settings they must be wary of disturbing the other members of the community (plants and animals) with which they are interacting. However, pick-up sticks is about removing as many of the sticks as possible without disturbing the others, until there are no sticks left. In ecotourism, indeed in all forms of land use, we must learn to recognise when enough is enough.

Towards the end of 1997 I had the opportunity to sit in on the deliberations of a provincial meeting on the development of the ecotourism product along the Trans-Canada Highway. The feeling was that rather than travel through a province, and hence view the highway simply as a medium of transportation, tourists should instead be encouraged to stop to enjoy some of the wonderful ecotourism sites along the way. This association of interested people grew as a consequence of the Trans-Canada Highway ecotours developed by Environment Canada in the 1970s and 1980s to allow people travelling on the highway (see Chapter 2) to enjoy natural attractions of various regions across Canada. However, whereas the Environment Canada publications were not developed with a profit motive in mind – they were simply a programme designed to strengthen the human–land relationship in Canadians and international travellers – the more recent meetings on the Trans-Canada Highway took on a very different form. There was an overriding emphasis on measuring success only in terms of the economic return from efforts put into marketing the product. There is no question that as the world's largest industry, tourism is very much about the profit motive, and, as outlined earlier, ecotourism happens to be the fastest-growing segment of the industry. In this meeting, however, there were decidedly few who could see the tourist as more than a monetary symbol. It made me wonder whether anything would be gained or accomplished if a person, visiting any given eco-attraction along the stretch of highway, did not spend a dime at a Trans-Canada ecotour site. So, what determines success of ecotourism ventures? People making money, or tourists gaining rewarding experiences (see the discussion in Chapter 7 on programme planning in ecotourism), or, indeed, both? Does the market really reflect the numbers of ecotourists reported these days, or is it simply the industry manufacturing and repackaging the product into ecotourism-like experiences for a public that is less than devoted to this form of travel? The reader will not find the answer to this question here, but nevertheless it is a question that needs to be addressed. Ecotourism is far too 'sexy' these days to be ignored by an industry bent on profit.

I thought again of ecolabelling when the chair of the meeting suggested that we 'take a broader look at ecotourism', including any cultural and natural attractions. What other types of attractions are there? Isn't it the same as the example of helicopter ecotourism in Hollywood mentioned in Chapter 5? But what about the differences between our helicopter in Hollywood and our hard-core ecotourist flying to Costa Rica? Which is more intrusive? There are those who argue that the former would in fact be less ecologically disruptive than having people traipsing through sensitive rainforests; after all, those in the helicopter aren't touching the ground. But yet the helicopter seems fundamentally wrong, for one reason or another. The answer could be that in order for the experience to occur, the ecotourist must have some tangible, concrete connection to the natural world. This means seeing it, feeling it, touching it – sensing it. All involve a type of learning that cannot occur in the helicopter; hence there is not the opportunity of getting 'into' nature. Also, is it the helicopter itself that is the main attraction, or the content of the experience (i.e. the interpretation of Hollywood)? Finally, how does it contribute to conservation? Would it be any more acceptable to take a helicopter flight over the Swiss Alps or over the Blue Mountains of Australia? Perhaps. But if helicopter flights over Hollywood become ecotourism, then anything can become ecotourism. The vultures are circling.

I have often contemplated the relationship that exists between tourism and recreation. Pleasure tourism falls within the realm of leisure and recreation. It is therefore a form of recreation and, like many of these forms, it involves spending money. However, recreation is also much more. It often brings out the best in community development and advocates accessibility in allowing others who are less fortunate to participate. Tourism, on the other hand, is elitist – not everyone can afford to travel – and is one of the best examples of the division between the haves and have-nots. The spirit of the provision of recreation services in the public and not-for-profit sectors has much more influence in the planning and programming of events than it does in tourism. There is the sense in tourism that events (tours) can be planned only with the profit motive in hand. However, will there ever be a day when tourism exists as more than just a vehicle for the private sector to make money, at the expense of accessibility? In recreation the experience occurs through the development of programmes (at least those that are not in the field of marketing); in tourism, we call the programme a product (see Fennell 2002b). The two have fundamentally different connotations. While in the future we could look for answers to this dilemma in the not-for-profit sector – which is more likely to have the interests of nature more firmly entrenched in its camp – many such agencies also adhere to a strong profit philosophy, as identified in Chapter 6.

Like the tourism industry in general, the fate of ecotourism will be controlled, to a certain degree, by unfettered demand. The question which is most pressing, though, is to what extent the international community is willing to let this happen. In simpler terms, will we let ecotourists and the providers of ecotourism services determine the fate of the industry, or rather will the industry require the intervention of other local, regional, national and international bodies to establish direction? There appears to be growing support for the latter, i.e. that regulation within the sustainable tourism and ecotourism industry is the way of the future (see Chapter 6). While in many cases this is unpalatable to those working in the industry, again we must look beyond satisfying strictly individual and economic needs to an approach that is sensitive to the needs of the resource base and local people first. To this end, the following list provides some direction in attaining more of the socio-ecological ends of a principled ecotourism culture:

1 Optimise the number of operators and tourists within ecotour regions in line with the capacity of the environment to absorb the impacts of ecotourism. This implies that significantly more research needs to be done to understand the relationship that exists between tourists, local people, the industry, government, and the resource base.

2 Conduct further research on the business philosophies of ecotour operators in understanding how they adhere to regional land use strategies. Allow only operators with proven track records to conduct tours. This entails a significant degree of monitoring to evaluate those that comply and those that do not.

3 Insist that operators maintain a specified level of competence, through accreditation, certification, or other manners which demonstrate competence. Insist also that evaluations be conducted by ecotourists and that these are subject to review by regional review committees (e.g. evaluations to be mailed not to the operator but rather to the committee).

4 Insist that governments and industry work together to institute fair and equitable guidelines to regulate and/or guide the ecotourism industry.

5 Research supports the notion that the development of an ecotourism industry should not be imposed on a community, but rather should start from the ground up. Such an approach emphasises the importance of decision-making from within rather than from outside. External influences (experts) should be encouraged, but not at the expense of the community's integrity.

6 Insist that international agencies be encouraged to coordinate research and to provide needed strength in calling attention to situations where the industry will potentially have or has had a significantly negative impact on the resources and people of a region.

7 Establish stronger linkages between educational institutions and local people by providing such people with the tools (e.g. diplomas, degrees) to work in the industry. These programmes need to be developed in association with the needs of local people, not necessarily the needs of the institutions.

8 Work to provide a definition of ecotourism and ecotourists which is acceptable to the international community. At present there is little consensus among practitioners and researchers, which is hampering the progress of research in the field. The foundation of ecotourism seems to be built on a few important variables, including a nature base, environmental education, minimal impact management, conservation and local control (the last two variables relating to sustainability), as outlined recently at the World Ecotourism Summit, in Quebec City, Canada (May 2002).

9 Develop a template to determine the percentage of ecotourism revenues that should go to local populations, and into the resource base of the host region.

10 Place more emphasis on addressing the critical absence of focus in ecotourism research, including research theory and methodologies. Included here would be the need to collect data that demonstrate to government and other decision-makers the benefits of eco-tourism, especially in the face of other, more disruptive forms of land use. This point also includes the need for more research funding in many countries around the world.

11 Related to the above is the notion that a multidisciplinary approach to ecotourism is greatly required if we are to expand our horizons. For example, ethics provides a bountiful crop of research which would allow research and decision-makers to examine morally good and bad decisions.

12 Examine international models where trusts have been established from tourism or other types of developments. The purpose of such trusts would be to accumulate money from tourism or other land uses, over time, so that in the event of the demise of the development or industry, the region would be able to maintain financial viability through investment in other industries. A case in point is Shetland, Scotland, which was able to build up a substantial trust from the oil industry of the North Atlantic. In the oil industry's absence, Shetland has been able to invest in other business ventures in balancing its economy.

As regards points 6 and 8 above, the recent International Year of Ecotourism in 2002 provided the first truly international collaborative effort to develop guidelines for ecotourism. The Quebec Summit represented the culmination of over twenty preparatory meetings around the world during 2002, and involved over 3,000 participants. Important in their deliberations was the involvement of many different stakeholder groups who shared their vision about ecotourism and their thoughts for a more ethically based ecotourism industry. The Quebec City Declaration on Ecotourism is presented in its entirety in Appendix 1, demonstrating the involvement of governments, industry operators, trade associations, non-governmental organisations, academics, consultants, intergovernmental organisations, communities and indigenous groups, and the responsibilities that each has for the future of ecotourism (see Appendix 1).

As discussed earlier, the responsibility for the planning and implementation of sound, ethical ecotourism programmes should fall on the shoulders of regional and national authorities. However, both tourists and operators must also accept a great deal of this responsibility. Accordingly, from the tourist's perspective, he or she must demand certain basic requirements that will ensure a high-quality and ethical experience. These requirements include the following:

1 Are local people employed in the operation of the business?
2 Are local people in middle or upper management of the firm?

3  Is respect shown for indigenous/local people?
4  Is respect shown for the wildlife of the region, and what precautions are taken to ensure this?
5  How is the operation economically beneficial to local people?
6  Does the operation use green hotels, restaurants, transportation, etc.? Question the type of sewage and waste management.
7  The degree of environmental education delivered in the product?
8  Number of participants in the tour group?
9  Level of experience and education of field guides?
10  Language barriers to communication, if applicable?
11  Has the tour operator extensive knowledge of the trip (i.e. has the operator been to the site before, and how many times)?
12  How much pre-trip and trip information is provided to ecotourists?
13  Contribution to conservation, either financially or physically (planting trees or removing garbage)?
14  How in touch is the operator with local land use and park management policies (guidelines, regulations)?
15  Does the operator follow a code of ethics?

In summary, the quotation from Burroughs at the outset of this chapter makes reference to the fact that we live in a superficial and valueless world. As a major global economic force, tourism is caught in the mainstream of a complex civilisation that tends to emphasise material ends over other, more virtuous ends. As part of the tourism industry, ecotourism will have to struggle to identify itself as part of either the conventional front or the alternative front. The coming years will thus be important in defining just what 'alternative' means, and the role that ecotourism plays in building a better alternative. If it is more about false burdens, vanities, and artificial wants, then ecotourism, I submit, will not be successful in its true meaning. If, on the other hand, it demands a values-based approach in philosophy and application, then indeed it will have something to contribute to human–human and human–environment relationships and, in doing so, may act as a model for other forms of development – tourism and non-tourism – in a very complex world.

## Summary questions

1  Which stakeholder group is observed to be most at fault for a lack of ethical leadership in the ecotourism industry?
2  What are some of the main arguments by critics of the ecotourism industry?
3  Why is it dangerous to measure success of an enterprise (like ecotourism) in economic terms only?
4  How is it that a focus on the philosophy of recreation service provision (the importance of provision) might temper the ill effects of the ecotourism industry (too much of a focus on material ends)?

# Appendix 1
# Québec City declaration
# on ecotourism

In the framework of the UN International Year of Ecotourism, 2002, under the aegis of the United Nations Environment Programme (UNEP) and the World Tourism Organization (WTO), over one thousand participants coming from 132 countries, from the public, private and non-governmental sectors met at the World Ecotourism Summit, hosted in Québec City, Canada, by Tourisme Québec and the Canadian Tourism Commission, between 19 and 22 May 2002.

The Québec Summit represented the culmination of 18 preparatory meetings held in 2001 and 2002, involving over 3,000 representatives from national and local governments including the tourism, environment and other administrations, private ecotourism businesses and their trade associations, non-governmental organizations, academic institutions and consultants, intergovernmental organizations, and indigenous and local communities.

This document takes into account the preparatory process, as well as the discussions held during the Summit. Although it is the result of a multistakeholder dialogue, it is not a negotiated document. Its main purpose is the setting of a preliminary agenda and a set of recommendations for the development of ecotourism activities in the context of sustainable development.

The participants at the Summit acknowledge the World Summit on Sustainable Development (WSSD) in Johannesburg, August/September 2002, as the ground-setting event for international policy in the next 10 years, and emphasize that, as a leading industry, the sustainability of tourism should be a priority at WSSD due to its potential contribution to poverty alleviation and environmental protection in critically endangered ecosystems. Participants therefore request the UN, its organizations and member governments represented at this Summit to disseminate the following Declaration and other results from the World Ecotourism Summit at the WSSD.

The participants to the World Ecotourism Summit, aware of the limitations of this consultative process to incorporate the input of the large variety of ecotourism stakeholders, particularly non-governmental organizations (NGOs) and local and indigenous communities,

*Acknowledge* that tourism has significant and complex social, economic and environmental implications,

*Consider* the growing interest of people in traveling to natural areas,

*Emphasize* that ecotourism should contribute to make the overall tourism industry more sustainable, by increasing economic benefits for host communities, actively contributing to the conservation of natural resources and the cultural integrity of host communities, and by increasing awareness of travelers towards the conservation of natural and cultural heritage,

*Recognize* the cultural diversity associated with natural areas, particularly because of the historical presence of local communities, of which some have maintained their traditional knowledge, uses and practices many of which have proven to be sustainable over the centuries,

*Reiterate* that funding for the conservation and management of biodiverse and culturally rich protected areas has been documented to be inadequate worldwide,

*Recognize* that sustainable tourism can be a leading source of revenue for protected areas,

*Recognize further* that many of these areas are home to rural peoples often living in poverty, who frequently lack adequate health care, education facilities, communications systems, and other infrastructure required for genuine development opportunity,

*Affirm* that different forms of tourism, especially ecotourism, if managed in a sustainable manner can represent a valuable economic opportunity for local populations and their cultures and for the conservation and sustainable use of nature for future generations,

*Emphasize* that at the same time, wherever and whenever tourism in natural and rural areas is not properly planned, developed and managed, it contributes to the deterioration of natural landscapes, threats to wildlife and biodiversity, poor water quality, poverty, displacement of indigenous and local communities, and the erosion of cultural traditions,

*Acknowledge* that ecotourism must recognize and respect the land rights of indigenous and local communities, including their protected, sensitive and sacred sites,

*Stress* that to achieve equitable social, economic and environmental benefits from ecotourism and other forms of tourism in natural areas, and to minimize or avoid potential negative impacts, participative planning mechanisms are needed that allow local and indigenous communities, in a transparent way, to define and regulate the use of their areas at the local level, including the right to opt out of tourism development,

*Note* that small and micro businesses seeking to meet social and environmental objectives are often operating in a development climate that does not provide suitable financial and marketing support for this specialized new market, and that to achieve this goal further understanding of the ecotourism market will be required through market research at the destination level, specialized credit instruments for tourism businesses, grants for external costs, incentives for the use of sustainable energy and innovative technical solutions, and an emphasis on developing skills not only in business but within government and those seeking to support business solutions,

In light of the above, the participants of the Summit produced a series of recommendations to governments, the private sector, non-governmental organizations, community-based associations, academic and research institutions, inter-governmental organizations, international financial institutions, development assistance agencies, and indigenous and local communities, presented in an annex to this Declaration.

Québec City, Canada, 22 May 2002
Annex I — Recommendations of the World Ecotourism Summit, Québec City, May 19 to 22, 2002.
The participants to the World Ecotourism Summit, having met in Québec City, from 19 to 22 May 2002, propose the following recommendations:

## A. To Governments

1. *formulate* national, regional and local ecotourism policies and development strategies that are consistent with the overall objectives of sustainable development, and to do so through a wide consultation process with those who are likely to become involved in, affect, or be affected by ecotourism activities. Furthermore, the principles that apply to ecotourism should be broadened out to cover the entire tourism sector;

2. In conjunction with local communities, the private sector, NGOs and all ecotourism stakeholders, *guarantee* the protection of nature, local cultures and specially traditional knowledge and genetic resources;

3. *ensure* the involvement, appropriate participation and necessary coordination of all the relevant public institutions at the national, provincial and local level, (including the establishment of inter-ministerial working groups as appropriate) at different stages in the ecotourism process, while at the same time opening and facilitating the participation of other stakeholders in ecotourism-related decisions. Furthermore, adequate budgetary mechanisms and appropriate legislative frameworks *be set up* to allow implementation of the objectives and goals set up by these multistakeholder bodies;

4. *include* in the above framework the necessary regulatory and monitoring mechanisms at the national, regional and local levels, including objective sustainability indicators jointly agreed with all stakeholders and environmental impact assessment studies, to prevent or minimize the occurrence of negative impacts upon communities or the natural environment. Monitoring results should be made available to the general public, since this information will allow tourists to choose an operator who adopts ecotourism principles over one who does not;

5. *develop* the local and municipal capacity to implement growth management tools such as zoning, and participatory land-use planning not only in protected areas but in buffer zones and other ecotourism development zones;

6. *use* internationally approved and reviewed guidelines to develop certification schemes, ecolabels and other voluntary initiatives geared towards sustainability in ecotourism, encouraging private operators to join such schemes and promoting their recognition by consumers. However, certification systems should reflect regional and sub-regional criteria and build capacity and provide financial support to make these schemes accessible to small and medium enterprises (SMEs). A regulatory framework is needed for such schemes to fulfill their mission;

7. *ensure* the provision of technical, financial and human resources development support to micro, small and medium-sized firms, which are the core of ecotourism, with a view to enable them to start, grow and develop their businesses in a sustainable manner. Similarly, that appropriate infrastructure is established in areas with ecotourism potential to stimulate the emergence of local enterprises.

8. *define* appropriate policies, management plans, and interpretation programs for visitors, and to earmark adequate sources of funding for protected natural areas to manage rapidly growing visitor numbers and protect vulnerable ecosystems, and effectively prevent the use of conservation hotspots. Such plans should include clear norms, direct and indirect management strategies, and regulations with the funds to ensure monitoring of social and environmental impacts for all ecotourism businesses operating in the area, as well as for tourists wishing to visit them;

9. *include* micro, small and medium-sized ecotourism companies, as well as community-based and NGO-based ecotourism operations in the overall promotional strategies and programmes carried out by the National Tourism Administration, both in the international and domestic markets;

10. *develop* regional networks and cooperation for promotion and marketing of ecotourism products at the international and national levels;

11. *provide* incentives to tourism operators (such as marketing and promotion advantages) for them to adopt ecotourism principles and make their operations more environmentally, socially and culturally responsible;

12. *ensure* that basic environmental and health standards are defined for all ecotourism development even in the most rural areas and in national and regional parks, that can play a pilot role. This should include aspects such as site selection, planning, design,

the treatment of solid waste, sewage, and the protection of watersheds, etc., and *ensure* also that ecotourism development strategies are not undertaken by governments without investment in sustainable infrastructure and the reinforcement of local/municipal capabilities to regulate and monitor such aspects;

13. *invest*, or support institutions that invest in research programmes on ecotourism and sustainable tourism. To institute baseline studies and surveys that record plant and animal life, with special attention to endangered species, as part of an environmental impact assessment (EIA) for any proposed ecotourism development;

14. *support* the further development of the international principles, guidelines and codes of ethics for sustainable tourism (e.g. such as those proposed by the Convention on Biological Diversity, UNEP, WTO) for the enhancement of international and national legal frameworks, policies and master plans to implement the concept of sustainable development into tourism;

15. *consider* as one option the reallocation of tenure and management of public lands, from extractive or intensive productive sectors to tourism combined with conservation, wherever this is likely to improve the net social, economic and environmental benefit for the community concerned;

16. *promote* and develop educational programmes addressed to children and young people to enhance awareness about nature conservation and sustainable use, local and indigenous cultures and their relationship with ecotourism;

17. *promote* collaboration between outbound tour operators and incoming operators and other service providers and NGOs at the destination to further educate tourists and influence their behaviour at destinations, especially those in developing countries.

## B. The private sector

18. *conceive, develop and conduct* their businesses minimizing negative effects on, and positively contributing to, the conservation of sensitive ecosystems and the environment in general, and directly benefiting local communities;

19. *bear* in mind that for ecotourism businesses to be sustainable, they need to be profitable for all stakeholders involved, including the projects' owners, investors, managers and employees, as well as the communities and the conservation organizations of natural areas where it takes place;

20. *adopt* a reliable certification or other systems of voluntary regulation, such as ecolabels, in order to demonstrate to their potential clients their adherence to sustainability principles and the soundness of the products and services they offer;

21. *cooperate* with governmental and non-governmental organizations in charge of protected natural areas and conservation of biodiversity, ensuring that ecotourism operations are practiced according to the management plans and other regulations prevailing in those areas, so as to minimize any negative impacts upon them while enhancing the quality of the tourism experience and contribute financially to the conservation of natural resources;

22. *make* increasing use of local materials and products, as well as local logistical and human resource inputs in their operations, in order to maintain the overall authenticity of the ecotourism product and increase the proportion of financial and other benefits that remain at the destination. To achieve this, private operators should invest in the training of the local workforce;

23. *ensure* that the supply chain used in building up an ecotourism operation is thoroughly sustainable and consistent with the level of sustainability aimed at in the final product or service to be offered to the customer;

24. *work* actively with indigenous leadership to ensure that indigenous cultures and communities are depicted accurately and with respect, and that their staff and guests are well and accurately informed regarding local indigenous sites, customs and history;

25. *promote* among their clients, the tourists, a more ethical behavior vis-à-vis the ecotourism destinations visited, providing environmental education to travelers, professionals and fostering inter-cultural understanding, as well as encouraging voluntary contributions to support local community or conservation initiatives;

26. *diversify* their offer by developing a wide range of tourist activities at a given destination and extending their operation to different destinations in order to spread the potential benefits of ecotourism and to avoid overcrowding some selected ecotourism sites, thus threatening their long-term sustainability. In this regard, private operators are urged to respect, and contribute to, established visitor impact management systems of ecotourism destinations;

27. *create and develop* funding mechanisms for the operation of business associations or cooperatives that can assist with ecotourism training, marketing, product development, research and financing;

28. In relation to the above points, *formulate* and *implement* company policies for sustainable tourism with a view to applying them in each part of the ecotourism operation.

## C. Non-governmental organizations, community-based associations, academic and research institutions

29. *provide* technical, financial, educational, capacity building and other support to ecotourism destinations, host community organizations, small businesses and the corresponding local authorities in order to ensure that appropriate policies, development and management guidelines, and monitoring mechanisms are being applied towards sustainability;

30. *monitor* and conduct research on the actual impacts of ecotourism activities upon ecosystems, biodiversity, local indigenous cultures and the socio-economic fabric of the ecotourism destinations;

31. *cooperate* with public and private organizations ensuring that the data and information generated through research is channeled to support decision-making processes in ecotourism development and management;

32. *cooperate* with research institutions to develop the most adequate and practical solutions to ecotourism development issues.

## D. Inter-governmental organizations, international financial institutions and development assistance agencies

33. *develop* and *assist* in the implementation of national and local policy and planning guidelines and evaluation frameworks for ecotourism and its relationships with biodiversity conservation, socio-economic development, respect of human rights, poverty alleviation, nature conservation and other objectives of sustainable development, and to intensify the transfer of such know-how to all countries. Special attention should be paid to countries in a developing stage or least developed status, to small island developing states and to countries with mountain areas, regarding that 2002 is also designated as the International Year of Mountains by the UN;

34. *build capacity* for regional, national and local organizations for the formulation and application of ecotourism policies and plans, based on international guidelines;

35. *develop* international standards and financial mechanisms for ecotourism certification systems that takes into account needs of small and medium enterprises and facilitates their access to those procedures;

36. *incorporate* multistakeholder dialogue processes into policies, guidelines and projects at the global, regional and national levels for the exchange of experiences between countries and sectors involved in ecotourism;

37. *strengthen* their efforts in identifying the factors that determine the success or failure of ecotourism ventures throughout the world, in order to transfer such experiences and best practices to other nations, by means of publications, field missions, training seminars and technical assistance projects; UNEP and WTO should continue this international dialogue after the Summit on sustainable ecotourism issues, for example by conducting periodical evaluations of ecotourism development through international and regional forums.

38. *adapt* as necessary their financial facilities and lending conditions and procedures to suit the needs of micro-, small- and medium-sized ecotourism firms that are the core of this industry, as a condition to ensure its long term economic sustainability;

39. *develop* the internal human resource capacity to support sustainable tourism and ecotourism as a development sub-sector in itself and to ensure that internal expertise, research, and documentation are in place to oversee the use of ecotourism as a sustainable development tool.

## E. Local communities and municipal organizations

40. As part of a community vision for development, that may include ecotourism, *define and implement* a strategy for improving collective benefits for the community through ecotourism development including human, physical, financial, and social capital development, and improved access to technical information;

41. strengthen, nurture and encourage the community's ability to maintain and use traditional skills that are relevant to ecotourism, particularly home-based arts and crafts, agricultural produce, traditional housing and landscaping that use local natural resources in a sustainable manner.

Accessed 18 October 2002
http://www.travelmole.com/cgi-bin/item.cgi?id=81853

# Bibliography

Adams, S.M. (1983) 'Public/private sector relations', *The Bureaucrat*, Spring: 7 10.

Akehurst, G. (1992) 'European Community tourism policy', in P. Johnson and B. Thomas (eds) *Perspectives on Tourism Policy*, London: Mansell.

Alderman, C. (1992) 'The economics and the role of privately owned lands used for nature tourism, education and conservation', Paper presented at the 4th World Congress on Parks and Protected Areas, Caracas, February.

Ammon, R. (1997) 'Risk management process', in D.J. Cotten and T.J. Wilde (eds) *Sport Law for Sport Managers*, Dubuque, IA: Kendall/Hunt.

Andersen, D.L. (1993) 'A window to the natural world: the design of ecotourism facilities', in K. Lindberg and D.E. Hawkins (eds) *Ecotourism: A Guide for Planners and Managers*, North Bennington, VT: The Ecotourism Society.

—— (1994) 'Ecotourism destinations: conservation from the beginning', *Trends* 31(2): 31–38.

Anderson/Fast (1996a) *Ecotourism in Saskatchewan: Primary Research*, Saskatoon, Saskatchewan: Anderson/Fast.

Anderson/Fast (1996b) *Ecotourism in Saskatchewan: State of the Resource (Report 1)*, Saskatoon, Saskatchewan: Anderson/Fast.

Applegate, J.E. and Clark, K.E. (1987) 'Satisfaction levels of birdwatchers: an observation on the consumptive– nonconsumptive continuum', *Leisure Sciences* 9: 129–134.

ARA Consultants (1994) *Ecotourism – Nature, Adventure/Culture: Alberta and British Columbia Market Demand Assessment*, Vancouver: ARA Consultants.

Aridjis, H. (2000) 'Flight of kings', *The Amicus Journal* (22)2: 26–29.

Arlen, C. (1995) 'Ecotour, hold the eco', *US News and World Report*, May 29.

Association for Experiential Education (1993) *Manual of Accreditation Standards for Adventure Programs*, Boulder, CO: AEE.

Atwood, M. (1984) 'The Galápagos haven't changed much in a million years. Now the tourists are coming', *Quest* (April): 37–45.

Ayala, H. (1996) 'Resort ecotourism: a paradigm for the 21st century', *Cornell Hotel and Restaurant Administration Quarterly* 37(5): 46–53.

Aylward, B., Allen, K., Echeverría, J., and Tosi, J. (1996) 'Sustainable ecotourism in Costa Rica: the Monteverde Cloud Forest Preserve', *Biodiversity and Conservation* 5(3): 315–343.

Bachert, D.W. (1990) 'Wilderness education: a holistic model', in A.T. Easley, J.F. Passineau, and B.L. Driver (eds) *The Use of Wilderness for Personal Growth, Therapy, and Education*, General Technical Report RM-193, Fort Collins, CO: USDA Forest Service.

Backman, S., Petrick, J., and Wright, B.A. (2001) 'Management tools and techniques: an integrated approach to planning', in D.B. Weaver (ed.) *The Encyclopedia of Ecotourism*, Wallingford, Oxon.: CAB International.

Baker, L., Woenne-Green, S., and the Mutitjulu Community (1992) 'The role of aboriginal ecological knowledge in ecosystem management', in J. Birckhead, I. De Lacy, and L. Smith (eds) *Aboriginal Involvement in Parks and Protected Areas*, Canberra: Aboriginal Studies Press.

Ballantine, J. and Eagles, P.F.J. (1994) 'Defining the Canadian ecotourist', *Journal of Sustainable Tourism* 2(4): 210–214.

Barbier, E.B. (1987) 'The concept of sustainable economic development', *Environmental Conservation* 14(2): 101–109.

Barkin, D. (2000) 'The economic impacts of ecotourism: conflicts and solutions in Highland Mexico', in P.M. Goode, M.F. Price, and F.M. Zimmermann (eds) *Tourism and Development in Mountain Regions*, London: CABI Publishing.

Barnwell, K.E. and Thomas, M.P. (1995) 'Tourism and conservation in Dalyan, Turkey: with reference to the loggerhead turtle (*Caretta caretta*) and the Euphrates turtle (*Trionyx triunguis*)', *Environmental Education and Information* 14(1): 19–30.

Barrow, G. (1994) 'Interpretive planning: more to it than meets the eye', *Environmental Interpretation* 9(2): 5–7.

Bassin, Z., Breault, M., Flemming, J., Foell, S., Neufeld, J., and Priest, S. (1992) 'AEE organizational membership preference for program accreditation', *Journal of Experiential Education* 15(2): 21–27.

Battisti, G. (1982) 'Central places and peripheral regions in the formulation of a theory on tourist space', in T.V. Singh and Jagdish Kaur (eds) *Studies in Tourism Wildlife Parks Conservation*, New Delhi: Metropolitan.

Bauer, T. and Dowling, R. (2003) 'Ecotourism policies and issues in Antarctica', in D.A. Fennell and R.K. Dowling (eds) *Ecotourism: Policy and Strategy Issues*, Wallingford, Oxon.: CAB International.

Belasco, J.A. and Stayer, R.C. (1993) *Flight of the Buffalo*, New York: Warner.

Beres, L. (1986) 'Contracting out! The pros and cons', Paper presented at the CPRA National Conference, Edmonton, Alberta.

Berkes, F. (1984) 'Competition between commercial and sport fisherman: an ecological analysis', *Human Ecology* 12(4): 413–429.

Black, R., Ham, S., and Weiler, B. (2000) *Ecotour Guide Training in Less Developed Countries: Some Research Directions for the 21st Century*, Working Paper 07/00, Faculty of Business and Economics, Monash University, Caufield East, Victoria, Australia.

Blackstone Corporation (1996) *Developing an Urban Ecotourism Strategy for Metro Toronto: A Feasibility Assessment for the Green Tourism Partnership*, Toronto, Canada.

Blamey, R.K. (1995) *The Nature of Ecotourism*, Occasional Paper No. 21, Canberra, ACT: Bureau of Tourism Research.

Blane, J. and Jackson, R. (1994) 'The impact of ecotourism boats on the St Lawrence beluga whales', *Environmental Conservation* 21(3): 267–269.

Blangy, S. and Nielson, T. (1993) 'Ecotourism and minimum impact policy', *Annals of Tourism Research* 20(2): 357–360.

Bodansky, D. (1994) 'The precautionary prnciple in US environmental law', in T. O'Riordan and J. Cameron (eds) *Interpreting the Precautionary Principle*, London: Earthscan.

Bonham, C. and Mak, J. (1996) 'Private versus public financing of state destination promotion', *Journal of Travel Research* 35(2): 3–10.

Boo, E. (1990) *Ecotourism: The Potentials and Pitfalls*, Washington, DC: World Wildlife Fund.

—— (1992) *The Ecotourism Boom: Planning for Development and Management*, Washington, DC: Wildlands and Human Needs, World Wildlife Fund.

Borg, W.R. and Gall, M.D. (1989) *Educational Research: An Introduction* (5th edn), New York: Longman.

Bottrill, C.G. and Pearce, D.G. (1995) 'Ecotourism: towards a key elements approach to operationalising the concept', *Journal of Sustainable Tourism* 3(1): 45–54.

Bowie, N.E. (1986) 'Business ethics', in J.P. DeMarco and R.M. Fox (eds) *New Directions in Ethics: The Challenge of Applied Ethics*, New York: Routledge & Kegan Paul.

Bowler, P.J. (1993) *The Norton History of the Environmental Sciences*, New York: W.W. Norton.

Boyd, S.W. (1991) 'Towards a typology of tourism: setting and experience', Paper presented at the Annual Meeting of the Association of American Geographers, Ohio State University, Youngstown, Ohio, November 1–2.

Boyd, S.W. and Butler, R.W. (1996) 'Managing ecotourism: an opportunity spectrum approach', *Tourism Management* 17(8): 557–566.

Brandon, K. (1996) *Ecotourism and Conservation: A Review of Key Issues*, Environment Department Paper No. 23, Washington, DC: The World Bank.

Bright, A.D., Fishbein, M., Manfredo, M., and Bath, A. (1993) 'Application of the theory of reasoned action to the National Park Service's controlled burn', *Journal of Leisure Research* 25: 263–280.

British Columbia Ministry of Development, Industry, and Trade (1991) *Developing a Code of Ethics: British Columbia's Tourism Industry*, Victoria, British Columbia: Ministry of Development, Trade, and Tourism.

Britton, R.A. (1977) 'Making tourism more supportive of small-state development: the case of St. Vincent', *Annals of Tourism Research* 4(5): 268–278.

Britton, S.G. (1982) 'The political economy of tourism in the Third World', *Annals of Tourism Research* 9(3): 331–358.

Brock, T.D. (1979) *Biology of Microorganisms* (3rd edn), Englewood Cliffs, NJ: Prentice-Hall Inc..

Brody, B. (1983) *Ethics and its Applications*, New York: Harcourt Brace Jovanovich.

Brookfield, H. (1975) *Interdependent Development*, London: Methuen.

Buckley, R. (1994) 'A framework for ecotourism', *Annals of Tourism Research* 21(3): 661–665.

Buckley, R. (in press) 'Tourism ecocertification in the international year of ecotourism', *Journal of Ecotourism*.

Buckley, R. and Pannell, J. (1990) 'Environmental impacts of tourism and recreation in national parks and conservation reserves', *Journal of Tourism Studies* 1(1): 24–32.

Budowski, G. (1976) 'Tourism and environmental conservation: conflict, coexistence, or symbiosis', *Environmental Conservation* 3(1): 27–31.

Bujold, P. (1995) 'Community development – making a better home', *Voluntary Action News*, 5–8.

Bull, A. (1991) *The Economics of Travel and Tourism*, Melbourne: Pitman.

Burch, W.R. Jr (1988) 'Human ecology and environmental management', in J.K. Agee and D.R. Johnson (eds) *Ecosystem Management for Parks and Wilderness*, Seattle: University of Washington Press.

Burger, J., Gochfeld, M., and Niles, L.J. (1995) 'Ecotourism and birds in coastal New Jersey: contrasting responses to birds, tourists, and managers', *Environmental Conservation* 22(1): 56–65.

Burr, S.W. (1995) 'Sustainable tourism development and use: follies, foibles, and practical approaches', in S.F. McCool and A.E. Watson (eds) *Linking Tourism, the Environment, and Sustainability*, USDA Technical Report INT-GTR-323, Ogden, UT: US Department of Agriculture, Forest Service, Intermountain Research Station.

Busser, J.A. (1990) *Programming: For Employee Services and Recreation*, Champaign, IL: Sagamore.

Butler, R.W. (1980) 'The concept of tourist area cycle of evolution: implications for management of resources', *The Canadian Geographer* 24: 5–12.

—— (1985) 'Evolution of tourism in the Scottish Highlands', *Annals of Tourism Research* 12(3): 379–391.

—— (1990) 'Alternative tourism: pious hope or Trojan horse?' *Journal of Travel Research* 28(3): 40–45.

—— (1991) 'Tourism, environment, and sustainable development', *Environmental Conservation* 18(3): 201–209.

—— (1993) 'Integrating tourism and resource management: problems of complementarity', in M.E. Johnston and W. Haider (eds) *Communities, Resources and Tourism in the North*, Thunder Bay, Ontario: Centre for Northern Studies, Lakehead University.

Butler, R.W. and Fennell, D.A. (1994) 'The effects of North Sea oil development on the development of tourism', *Tourism Management* 15(5): 347–357.

Butler, R.W. and Waldbrook, L.A. (1991) 'A new planning tool: the tourism opportunity spectrum', *Journal of Tourism Studies* 2(1): 2–14.

Butler, R.W., Fennell, D.A., and Boyd, S.W. (1992) *The POLAR Model: A System for Managing the Recreational Capacity of Canadian Heritage Rivers*, Ottawa: Environment Canada.

Butynski, T. (1998) 'Is gorilla tourism sustainable', *Gorilla Journal* 16: 1–3.

Campbell, A., Devine, M., and Young, D. (1990) *A Sense of Mission*, London: Pitman Press.

Canada, Government (1990) *Canada's Green Plan*, Ottawa: Supply and Services Canada.

Canada, Parliament (1993) *Statutes of Canada Revised Loose Leaf Edition*, Chapter N-14, Ottawa: Supply and Services Canada.

Canadian Environmental Advisory Council (1991) *A Protected Areas Vision for Canada*, Ottawa: Minister of Supply and Services.

—— (1992) *Ecotourism in Canada*, Ottawa: Minister of Supply and Services.

Canadian Tourism Commission (1995) *Adventure Travel in Canada: An Overview of Product, Market and Business Potential*, Ottawa: Tourism Canada.

Canova, L. (1994) 'Tourism in the modern age', *All of Us* 15: (unknown).

Carpenter, G.M. and Howe, C.Z. (1985) *Programming Leisure Experiences: A Cyclical Approach*, Englewood Cliffs, NJ: Prentice-Hall.

Ceballos-Lascuráin, H. (1987) *Estudio de prefactibilidaad socioeconómica del turismo ecológico y anteproyecto asquitectónico y urbanístico del Centro de Turismo Ecológico de Sian Ka'an, Quintana Roo*. Study completed for SEDUE, Mexico.

—— (1996a) *Tourism, Ecotourism, and Protected Areas*, Gland, Switzerland: International Union for the Conservation of Nature and Natural Resources.

—— (1996b) 'Ecotourism or a second Cancún in Quintana Roo?', *The Ecotourism Society Newsletter*, third quarter: 1–3.

Center for Tourism Policy Studies (1994) *Ecotourism Opportunities for Hawaii's Visitor Industry*, Honolulu: Department of Business, Economic Development and Tourism.

Cerruti, J. (1964) 'The two Acapulcos', *National Geographic Magazine* 126(6): 848–878.

Chapin, M. (1990) 'The silent jungle: ecotourism among the Kuna Indians of Panama', *Cultural Survival Quarterly* 14(1): 42–45.

Chase, A. (1987) 'How to save our national parks', *Atlantic Monthly*, July: 35–44.

—— (1989) 'The Janzen heresy', *Condé Nast Traveler*, November: 122–127.

Chester, G. (1997) 'Australian ecotourism accreditation off and running', *Ecotourism Society Newsletter*, second quarter: 9, 11.

Chipeniuk, R. (1988) 'The vacant niche: an argument for the re-creation of a hunter–gatherer component in the ecosystems of northern national parks', *Environments* 20(1): 50–59.

Chodos, R. (1977) *The Caribbean Connection*, Toronto: James Lorimer.

Christaller, W. (1963) 'Some considerations of tourism location in Europe: the peripheral regions – underdeveloped countries – recreation areas', *Regional Science Association Papers* 6: 95–105.

Christensen, N.A. (1995) 'Sustainable community-based tourism and host quality of life', in S.F. McCool and A. Watson (eds) *Linking Tourism, the Environment, and Sustainability*, USDA Forest Service General Technical Report INT-GTR-323, pp. 63–68.

Christiansen, D.R. (1990) 'Adventure tourism', in J.C. Miles and S. Priest (eds) *Adventure Education*, State College, PA: Venture Publishing.

Chubb, M. and Chubb, H.R. (1981). *One-third of Our Time? An Introduction to Recreation Behavior and Resources*, New York: John Wiley.

Clawson, M. and Knetsch, J.L. (1966) *Economics of Outdoor Recreation*, Baltimore: Johns Hopkins University Press.

Clements, C.J., Schultz, J.H., and Lime, D.W. (1993) 'Recreation, tourism, and the local residents: partnership or co-existence?' *Journal of Park and Recreation Administration* 11(4): 78–91.

Coccossis, H. (1996) 'Tourism and sustainability: perspectives and implications', in G.K. Priestley, J.A. Edwards, and H. Coccossis (eds) *Sustainable Tourism? European Experiences*, Wallingford, Oxon.: CAB International.

Cockrell, D. and Detzel, D. (1985) 'Effects of outdoor leadership certification on safety, impacts, and program', *Trends* 22(3): 15–21.

Cohen, E. (1972) 'Toward a sociology of tourism', *Social Research* 39(1): 164–182.

—— (1978) 'The impact of tourism on the physical environment', *Annals of Tourism Research* 5(2): 215–237.

—— (1987) 'Alternative tourism – a critique', *Tourism Recreation Research* 12(2): 13–18.

Coleman, J.S. (1988) 'Social capital in the creation of human capital', *American Journal of Sociology* 84 (suppl.): S95–S119.

Colvin, J.G. (1994) 'Capirona: a model of indigenous ecotourism', Paper presented at the Second Global Conference: Building a Sustainable World through Tourism, Montreal, September.

Commonwealth of Australia (1997) *Ecotourism Snapshot: A Focus on Recent Market Research*, Canberra: Office on National Tourism.

Commonwealth Department of Tourism (1994) *National Ecotourism Strategy*, Canberra: Commonwealth of Australia.

Conservation International (1997) *Conservation International's Ecotourism Program Initiatives*, Washington, DC: CI.

Consulting and Audit Canada (1995) *What Tourism Managers Need to Know: A Practical Guide to the Development and Use of Indicators of Sustainable Tourism*, Ottawa: Consulting and Audit Canada, for the World Tourism Organization.

Cooper, C. (1995) 'Strategic planning for sustainable tourism: the case of the offshore islands of the UK', *Journal of Sustainable Tourism* 3(4): 191–209.

Cooper, C. and Jackson, S. (1989) 'Destination life cycle: the Isle of Man case study', *Annals of Tourism Research* 16(3): 377–398.

Coppock, J.T. (1982) 'Tourism and conservation', *Tourism Management* 3: 270–276.

Corbett, R. and Findlay, H. (1998) *Your Risk Management Program: A Handbook for Sport Organization*, Ottawa: Centre for Sport and Law.

Cordell, J. (1993) 'Who owns the land? Indigenous involvement in Australian protected areas', in E. Kemf (ed.) *The Law of the Mother: Protecting Indigenous People in Protected Areas*, San Francisco: Sierra Club.

Countryside Commission (1990) *National Parks in Focus*, Cheltenham: Countryside Commission.

Crabtree, A.E. and Black, R.S. (n.d.) *Ecoguide Program: Guide Workbook*, Brisbane: Ecotourism Association of Australia.

Crittendon, A. (1975) 'Tourism's terrible toll', *International Wildlife* 5(2): 4–12.

Csikszentmihalyi, M. (1975) *Beyond Boredom and Anxiety*, San Francisco, CA: Jossey-Bass, Inc.
—— (1990) *Flow: The Psychology of Optimal Experience*, New York: HarperCollins.

Csikszentmihalyi, M. and Csikszentmihalyi, I.S. (1990) 'Adventure and the flow experience', in J.C. Miles and S. Priest (eds) *Adventure Education*, State College, PA: Venture Publishing.

Deltabuit, M. and Pi-Sunyer, O. (1990) 'Tourism development in Quintana Roo, Mexico', *CS Quarterly* 14(1): 9–13.

D'Amore, L.J. (1993) 'A code of ethics and guidelines for socially and environmentally responsible tourism', *Journal of Travel Research* 31(3): 64–66.

Davey, S. (1993) 'Creative communities: planning and comanaging protected areas', in E. Kemf (ed.) *The Law of the Mother: Protecting Indigenous Peoples in Protected Areas*, San Francisco: Sierra Club Books.

Davidson, M. (1995) 'Community development', *Recreation Saskatchewan* 22: 5–6.

Davis, P.B. (1998) 'Beyond guidelines: a model for Antarctic tourism', *Annals of Tourism Research* 23(3): 546–553.

de Castro, F. and Bergossi, A. (1996) 'Fishing at Rio Grande (Brazil): ecological niche and competition', *Human Ecology* 24(3): 401–411.

DeGraaf, D.G., Jordan, D.J., and De Graaf, K.H. (1999) *Programming for Parks, TRecreation, and Leisure Services: A Servant Leadership Approach*, State College, PA: Venture Publishing.

De Groot, R.S. (1983) 'Tourism and conservation in the Galapagos Islands', *Biological Conservation* 26: 291–300.

De Lacy, T. (1992) 'Towards an aboriginal land management curriculum', in J. Birckhead, T. De Lacy, and L. Smith (eds) *Aboriginal Involvement in Parks and Protected Areas*, Canberra: Aboriginal Studies Press.

De Lacy, T. and Lockwood, M. (1992) 'Estimating the non-market values of nature conservation resources in Australia', Paper presented at the Fourth World Congress on Parks and Protected Areas, Caracas, February 10–21.

Dearden, P. (1991) 'Parks and protected areas', in B. Mitchell (ed.) *Resource Management and Development*, Toronto: Oxford University Press.

Dearden, P. and Rollins, R. (1993) 'The times they are a-changin'', in P. Dearden and R. Rollins (eds) *Parks and Protected Areas in Canada: Planning and Management*, Toronto: Oxford University Press.

Debbage, K.G. (1990) 'Oligopoly and the resort cycle in the Bahamas', *Annals of Tourism Research* 17(4): 513–527.

Deming, A. (1996) 'The edges of the civilized world: tourism and the hunger for wild places', *Orion* 15(2): 28–35.

Dernoi, L.A. (1981) 'Alternative tourism: towards a new style in North–South relations', *Tourism Management* 2: 253–264.

Devall, B. and Sessions, G. (1985) *Deep Ecology*, Salt Lake City: Gibbs Smith.

Diamantis, D. and Ladkin, A. (1999) '"Green" strategies in the tourism and hospitality industries', in F. Vellas and L. Becheral (eds) *The International Marketing of Travel and Tourism*, London: MacMillan Press.

Dieke, P. (2001) 'Kenya and South Africa', in D.B. Weaver (ed.) *The Encyclopedia of Ecotourism*, London: CAB International.

Dodds, R. and Joppe, M. (2001) 'Promoting urban green tourism: the development of the other map of Toronto', *Journal of Vacation Marketing* (7)3: 261–267.

Dowling, R.K. (1992) *The Ecoethics of Tourism: Guidelines for Developers, Operators and Tourists*, Canberra: Bureau of Tourism Research.

—— (1993) 'An environmentally based planning model for regional tourism development', *Journal of Sustainable Tourism* 1(1): 17–37.

Doxey, G.V. (1975) 'A causation theory of visitor–resident irritants: methodology and research inference', Paper presented at the TTRA Conference, San Diego, California, pp. 195–198.

Drake, S.P. (1991) 'Local participation in ecotourism projects', in T. Whelan (ed.) *Nature Tourism: Managing for the Environment*, Washington, DC: Island Press.

Duenkel, N. and Scott, H. (1994) 'Ecotourism's hidden potential – altering perceptions of reality', *Journal of Physical Education Recreation and Dance*, October: 40–44.

Dufrene, M. (1973) *The Phenomenology of Aesthetic Experience*, Evanston, IL: Northwestern University Press.

Dunlap, R.E. and Heffernan, R.B. (1975) 'Outdoor recreation and environmental concern: an empirical examination', *Rural Sociology* 40: 18–30.

Durst, P.B. and Ingram, C.D. (1988) 'Nature-oriented tourism promotion by developing countries', *Tourism Management* 9(1): 39–43.

Dyess, R. (1997) 'Adventure travel or ecotourism?', *Adventure Travel Business*, April: 2.

Eagles, P.F.J. (1992) 'The travel motivations of Canadian ecotourists', *Journal of Travel Research* 31(Fall): 3–7.

—— (1993) 'Parks legislation in Canada', in P. Dearden and R. Rollins (eds) *Parks and Protected Areas in Canada*, Toronto: Oxford University Press.

—— (1995) 'Tourism and Canadian parks: fiscal relationships', *Managing Leisure* 1: 16–27.

Eagles, P.F.J. and Wind, E. (1994) 'Canadian ecotours in 1992: a content analysis of advertising', *Journal of Applied Recreation Research* 19(1): 67–87.

Echeverría, J., Hanrahan, M., and Solorzano, R. (1995) 'Valuation of non-priced amenities provided by the biological resources within the Monteverde Cloud Forest Preserve, Costa Rica', *Ecological Economics* 13: 45–52.

Economic and Social Commission for Asia and the Pacific (ESCAP) (1978) 'The formulation of basic concepts and guidelines for preparation of tourism sub-regional master plans in the ESCAP region' (in Britton 1982).

*The Economist* (1991) 'Travel and tourism: the pleasure principle', *The Economist*, March: 3–22.

*The Economist* (2002) Accessed 14 June 2002 (http://0-ehostvgw20.epnet.com.brai . . . rHitNum =1&r1StartHit=1&delType=FT).

Ecotourism/Heritage Tourism Advisory Committee (1997) *Planning for the Florida of the Future*, Tallahassee, Florida: Government Document.

Ecotourism Society (1993) *Ecotourism Guidelines for Nature Tour Operators*, North Bennington, VT: The Ecotourism Society.

Ecotourism Society of Saskatchewan (2000) *Saskatchewan Ecotourism Accreditation System*, Regina, Saskatchewan.

Edginton, C.R., Compton, D.M., and Hanson, C.J. (1980) *Recreation and Leisure Programming: A Guide for the Professional*, Philadelphia, PA: Saunders.

Edginton, C.R., Hanson, C.J., Edginton, S.R., and Hudson, S.D. (1998) *Leisure Programming: A Service Centred and Benefits Approach*, Boston, MA: WCB/McGraw-Hill.

Eidsvik, H. (1983) 'Biosphere reserves/opportunities for cooperation: a global perspective', in R.C. Scace and C.J. Martinka (eds) *Towards the Biosphere Reserve: Exploring Relationships between Parks and Adjacent Lands*, Kalispell, MT: Department of the Interior, National Park Service.

—— (1993) 'Canada, conservation, and protected areas', in P. Dearden and R. Rollins (eds) *Parks and Protected Areas in Canada*, Toronto: Oxford University Press.

Elliot, R. (1991) 'Environmental ethics', in P. Singer (ed.) *A Companion to Ethics*, Oxford: Blackwell.

Environment Canada (1990) *National Parks System Plan*, Ottawa: Supply and Services Canada.

Erisman, H.M. (1983) 'Tourism and cultural dependency in the West Indies', *Annals of Tourism Research* 10(3): 337–361.

Ewert, A. (1985) 'Why people climb: the relationship of participant motives and experience level to mountaineering', *Journal of Leisure Research* 17(3): 241–250.

Ewert, A. and Shultis, J. (1997) 'Resource-based tourism: an emerging trend in tourism experiences', *Parks and Recreation*, September: 94–103.

Exmoor National Park (1990) Exmoor National Park Plan Second Review: Consultation Draft, Exmoor National Park Authority.

Farquharson, M. (1992) 'Ecotourism: a dream diluted', *Business Mexico* 2(6): 8–11.

Farr, H. and Rogers, A. (1994) 'Tourism and the environment on the Isles of Scilly: conflict and complementarity', *Landscape and Urban Planning* 29: 1–17.

Farrell, B.H. and McLellan, R.W. (1987) 'Tourism and physical environment research', *Annals of Tourism Research* 14(1): 1–16.

Farrell, P. and Lundegren, H.M. (1993) *The Process of Recreation Programming*, State College, PA: Venture Publishing.

Fayos-Solá, E. (1996) 'Tourism policy: a midsummer night's dream?' *Tourism Management* 17(6): 405–412.

Federation of Nature and National Parks of Europe (1993) *Loving Them to Death? Sustainable Tourism in Europe's Nature and National Parks*, Eupen, Belgium: Kliemo.

Fennell, D.A. (1990) 'A profile of ecotourists and the benefits derived from their experience: a Costa Rican case study', unpublished master's thesis, University of Waterloo, Waterloo, Ontario.

—— (1996) 'A tourist space–time budget in the Shetland Islands', *Annals of Tourism Research* 23(4): 811–829.

—— (1998) 'Ecotourism in Canada', *Annals of Tourism Research* 25(1): 231–234.

Fennell, David A. (2002a) 'The Canadian ecotourist in Costa Rica: ten years down the road', *International Journal of Sustainable Development* 5(3): 282–299.

—— (2002b) *Ecotourism Programme Planning*, Wallingford, Oxon.: CAB International.

Fennell, David A., Buckley, R., and Weaver, D.B. (2001) 'Policy and planning', in D.B. Weaver (ed.) *The Encyclopedia of Ecotourism*, Wallingford, Oxon.: CAB International.

Fennell, D.A. and Butler, R.W. (in press) 'A human ecological approach to tourism interactions', *Progress in Tourism and Hospitality Research*.

Fennell, D.A. and Dowling, R.K. (2003) 'Ecotourism policy and planning: stakeholders, management and governance', in D.A. Fennell and R.K. Dowling (eds) *Ecotourism Policy and Planning*, Wallingford, Oxon.: CAB International.

Fennell, D.A. and Eagles, P.F.J. (1990) 'Ecotourism in Costa Rica: a conceptual framework', *Journal of Park and Recreation Administration* 8(1): 23–34.

Fennell, D.A. and Malloy, D.C. (1995) 'Ethics and ecotourism: a comprehensive ethical model', *Journal of Applied Recreation Research* 20(3): 163–184.

—— (1997) 'Measuring the ethical nature of tourist operators: a comparison', World Congress and Exhibition on Ecotourism, Rio de Janeiro, Brazil, December 15–18, pp. 144–149.

Fennell, D.A. and Przeclawski, K. (2003) 'Tourists and host communities', in S. Singh, D.J. Timothy, and R.K. Dowling (eds) *Tourism in Destination Communities*, Wallingford, Oxon.: CAB International.

Fennell, D.A. and Smale, B.J.A. (1992) 'Ecotourism and natural resource protection: implications of an alternative form of tourism for host nations', *Tourism Recreation Research* 17(1): 21–32.

Fennell, J. (1989) 'Destination resorts', *Canadian Building* 39(5): 10–17.

Finch, R. and Elder, J. (1990) *The Norton Book of Nature Writing*, New York: W.W. Norton and Co.

Flinn, D. (1989) *Travellers in a Bygone Shetland*, Edinburgh: Scottish Academic Press.

Font, X. and Buckley, R. (2001) *Tourism Ecolabelling*, Wallingford, Oxon.: CAB International.

Ford, P. and Blanchard, J. (1993) *Leadership and Administration of Outdoor Pursuits*, State College, PA: Venture.

Forestell, P.H. (1993) 'If Leviathan has a face, does Gaia have a soul?: incorporating environmental education in marine eco-tourism programs', *Ocean and Coastal Management* 20(3): 267–282.

Forman, R.T.T. (1990) 'Ecologically sustainable landscapes: the role of spatial configuration', in I.S. Zonneveld and R.T.T. Forman (eds) *Changing Landscapes: An Ecological Perspective*, New York: Springer Verlag.

Forsyth, T. (1995) 'Business attitudes to sustainable tourism: self-regulation in the UK outgoing tourism industry', *Journal of Sustainable Tourism* 3(4): 210–231.

Fowkes, J. and Fowkes, S. (1991) 'Roles of private sector ecotourism in protected areas', *Parks* 2(3): 26–30.

Fox, A. (1996) 'Kakadu, tourism and the future', *Australian Natural History* 21(7): 266–271.

Fox, R.M. and DeMarco, J.P. (1986) 'The challenge of applied ethics', in R.M. Fox and J.P. DeMarco (eds) *New Directions in Ethics: The Challenge of Applied Ethics*, New York: Routledge & Kegan Paul.

Francis, G. (n.d.) 'Ecosystem management', Paper presented at the Tri-National Conference on the North American Experience Managing Transboundary Resources: The United States and the Boundary Commissions.

—— (1985) 'Biosphere reserves: innovations for cooperation in the search for sustainable development', *Environments* 17(3): 21–38.

Frangialli, F. (1997) Keynote address to the World Ecotour '97 Conference, Rio de Janeiro, December 15–18.

Freidmann, J. and Alonso, W. (eds) (1974) *Regional Development and Planning: A Reader*, Cambridge, MA: MIT Press.

Frost, J.E. and McCool, S.F. (1988) 'Can visitor regulations enhance recreational experiences?', *Environmental Management* 12(1): 5–9.

Gardner, J.E. (1993) 'Environmental non-government organizations (ENGOs) and sustainable development', in S. Lerner (ed.) *Environmental Stewardship: Studies in Active Earthkeeping*, Department of Geography Publication Series, No. 39, University of Waterloo, Waterloo, Ontario.

Gass, M. and Williamson, J. (1995) 'Accreditation for adventure programs', *JOPERD*, January, 22–27.

Genot, H. (1995) 'Voluntary environmental codes of conduct in the tourism sector', *Journal of Sustainable Tourism* 3(3): 166–172.

Getz, D. (1986) 'Models in tourism planning: towards integration of theory and practice', *Tourism Management* 7(1): 21–32.

—— (1990) *Festivals, Special Events, and Tourism*, New York: Van Nostrand Reinhold.

Giannecchini, J. (1993) 'Ecotourism: new partners, new relationships', *Conservation Biology* 7(2): 429–432.

Gilchrist, N.L. (1998) 'Winter hiking and Camping', in N.J. Dougherty IV (ed.) *Outdoor Recreation Safety*, Champaign, IL: Human Kinetics.

Gilmore, G.D. and Campbell, M.D. (1996) *Needs Assessment Strategies for Health Education and Health Promotion*, Madison, WI: WCB Brown and Benchmark.

Globe '90 (1990) *An Action Strategy for Sustainable Tourism Development*, Ottawa: Tourism Canada.

Goeldner, C.R., Ritchie, J.R.B, and McIntosh, R.W. (2000) *Tourism: Principles, Practices, Philosophies*, New York: John Wiley & Sons, Inc.

Goodall, B. (1994) 'Environmental auditing: current best practice (with special reference to British tourism firms)', in A.V. Seaton (ed.) *Tourism: The State of the Art*, Chichester: John Wiley.

Goodall, B. and Cater, E. (1996) 'Self regulation for sustainable tourism?', *Ecodecision* 20(Spring): 43–45.

Goodwin, H. (1995) 'Tourism and the environment', *Biologist* 42(3): 129–133.

—— (1996) 'In pursuit of ecotourism', *Biodiversity and Conservation* 5(3): 277–291.

Graburn, N. (1989) 'Tourism: the sacred journey', in V.L. Smith (ed.) *Hosts and Guests: The Anthropology of Tourism* (2nd edn), Philadelphia, PA: University of Pennsylvania Press.

Graefe, A.R., Vaske, J.J., and Kuss, F.R. (1984a) 'Resolved issues and remaining questions about social carrying capacity', *Leisure Sciences* 6(4): 497–507.

—— (1984b) 'Social carrying capacity: an integration and synthesis of twenty years of research', *Leisure Sciences* 6: 395–431.

Green, H. and Hunter, C. (1992) 'The environmental impact assessment of tourism development', in P. Johnson and B. Thomas (eds) *Perspectives on Tourism Policy*, London: Mansell.

Green Tourism Association (1999) 'Map Project Evaluation Report' and www.greentourism.on.ca (last accessed October 2001).

Grenier, A.A. (1998) *Ship-based Polar Tourism in the Northwest Passage: A Case Study*, Finland: University of Lapland.

Grenier, D., Kaae, B.C., Miller, M.L., and Mobley, R.W. (1993) 'Ecotourism, landscape architecture and urban planning', *Landscape and Urban Planning* 25(1–2): 1–16.

Grumbine, E. (1996) 'Beyond conservation and preservation in American environmental values', in B.L. Driver, D. Dustin, T. Baltic, G. Elsner, and G. Peterson (eds) *Nature and the Human Spirit: Toward an Expanded Land Management Ethic*, State College, PA: Venture.

Grupo de Trabalho Interministerial MICT/MMA (1994) *Diretrizes para uma Política Nacional de ecotourism/Coordenação de Silvio Magalhães Barros II e Denis Hamú m de La Penha*, Brasília: EMBRATUR.

Guignon, C. (1986) 'Existential ethics', in R.M. Fox and J.P. DeMarco (eds) *New Directions in Ethics: The Challenge of Applied Ethics*, New York: Routledge & Kegan Paul.

Gunn, C.A. (1972) *Vacationscape: Designing Tourist Regions*, University of Texas: Bureau of Business Research.

—— (1988) *Tourism Planning*, New York: Taylor & Francis.

Gurung, C. (1995) 'Ecolodges and their role in integrated conservation and development', Paper presented at the Second International Ecolodge Forum and Field Seminar, San José, Costa Rica, October 22–29.

Hadley, J. and Crow, P. (1995) 'Some guidelines for the architecture of ecotourism facilities', in *The Ecolodge Sourcebook for Planners and Managers*, North Bennington, VT: The Ecotourism Society.

Halbertsma, N.F. (1988) 'Proper management is a must', *Naturopa* 59: 23–24.

Hall, C.M. (1992) 'Adventure, sport and health tourism', in B. Weiler and C.M. Hall (eds) *Special Interest Tourism*, London: Belhaven Press.

Hall, C.M. and Jenkins, J. (1995) *Tourism and Public Policy*, London: Routledge.

Hall, C.M. and McArthur, S. (1998) *Integrated Heritage Management: Principles and Practice*, London: The Stationery Office.

Hall, C.M. and Page, S.J. (1999) *The Geography of Tourism and Recreation: Environment, Place and Space*, London: Routledge.

Hall, C.M. and Wouters, M. (1995) 'Issues in Antarctic tourism', in C.M. Hall and M.E. Johnston (eds) *Polar Tourism: Tourism in the Arctic and Antarctic Regions*, London: John Wiley.

Hall, S. (ed.) (1993) *Ethics in Hospitality Management: A Book of Readings*, East Lansing, MI: Educational Institute of the American Hotel and Motel Association.

Halpern, S. (1998) 'A fragile kingdom', *Audubon* (100)2: 37–45, 99–101.

Ham, S. and Weiler, B. (2000) *Six Principles for Tour Guide Training and Sustainable Development in Developing Countries*, Working Paper 102/00, Faculty of Business and Economics, Monash University, Caufield East, Victoria, Australia.

Hammitt, W.E. and Cole, D.C. (1987) *Wildland Recreation: Ecology and Management*, New York: John Wiley.

Harrington, I. (1971) 'The trouble with tourism unlimited', *New Statesman* 82: 176.

Harroun, L.A. (1994) *Potential Frameworks for Analysis of Ecological Impacts of Tourism in Developing Countries*, Washington, DC: World Wide Fund for Nature.

Harroy, J.P. (1974) 'A century in the growth in the "national park" concept throughout the world', in H. Elliot (ed.) *Second World Conference on National Parks*, Gland, Switzerland: International Union for the Conservation of Nature and Natural Resources.

Hawkes, S. and Williams, P. (1993) *The Greening of Tourism: From Principles to Practice*, Burnaby, British Columbia: Centre for Tourism Policy and Research, Simon Fraser University.

Hawkins, D. (1994) 'Ecotourism: opportunities for developing countries', in W. Theobald (ed.) *Global Tourism: The Next Decade*, Oxford: Butterworth.

Hays, S.P. (1959) *Conservation and the Gospel of Efficiency*, Cambridge, MA: Harvard University Press.

Hayward, S.J., Gomez, V.H., and Sterrer, W. (1981) *Bermuda's Delicate Balance*, Hamilton: Bermuda National Trust.

Haywood, M. (1986) 'Can the tourist area life cycle be made operational?', *Tourism Management* 7(3): 154–167.

Heald, D. (1984) 'Privatization: analysing its appeal and limitations', *Fiscal Studies* 5(1): 9–15.

Hendee, J.C., Stankey, G.H., and Lucas, R.C. (1990) *Wilderness Management*, Golden, CO: International Wilderness Leadership Foundation.

Henderson, N. (1992) 'Wilderness and the nature conservation ideal: Britain, Canada, and the United States contrasted', *Ambio* 21(6): 394–399.

Herfindahl, O. (1961) 'What is conservation', in O.C. Herfindahl (ed.) *Three Studies in Mineral Economics*, Washington, DC: Resources for the Future.

Hetzer, N.D. (1965) 'Environment, tourism, culture', *LINKS* (July); reprinted in *Ecosphere* (1970) 1(2): 1–3.

Higgins, B.R. (1996) 'The global structure of the nature tourism industry: ecotourists, tour operators, and local businesses', *Journal of Travel Research* 35(2): 11–18.

Higham, J. and Lück, M. (2002) 'Urban ecotourism: a contradiction in terms?', *Journal of Ecotourism* 1(1): 36–51.

Hills, T. and Lundgren, J. (1977) 'The impact of tourism in the Caribbean: a methodological study', *Annals of Tourism Research* 4(5): 248–267.

Hitchcock, R.K. (1993) 'Toward self-sufficiency', *Cultural Survival Quarterly* 17(2): 51–53.

Hjalager, A.-M. (1996) 'Tourism and the environment: the innovation connection', *Journal of Sustainable Tourism* 4(4): 201–218.

Hockings, M. (1994) 'A survey of the tour operator's role in marine park interpretation', *Journal of Tourism Studies* 5(1): 16–28.

Hoffman, W.M., Frcdcrisk, R.E., and Schwartz, M.S. (2001) *Business Ethics: Readings and Cases in Corporate Morality*, Boston, MA: McGraw-Hill.

Holloway, J.C. (1994) *The Business of Tourism* (4th edn), London: Pitman.

Honderich, T. (1995) *The Oxford Companion to Philosophy*, New York: Oxford University Press.

Honey, M. (1999) *Ecotourism and Sustainable Development: Who Owns Paradise?*, Washington, DC: Island Press.

Honey, M. and Rome, A. (2001) *Protecting Paradise: Certification Programs for Sustainable Tourism and Ecotourism*, Washington, DC: Institute for policy Studies.

Hope, K.R. (1980) 'The Caribbean tourism sector: recent performance and trends', *Tourism Management* 1(3): 175–183.

Hough, J.L. (1988) 'Obstacles to effective management of conflicts between national parks and surrounding human communities in developing countries', *Environmental Conservation* 15(2): 129–136.

Hovinen, G.R. (1981) 'The tourist cycle in Lancaster County, Pennsylvania', *Canadian Geographer* 25(3): 286–289.

Hughes, G. (1995) 'The cultural construction of sustainable tourism', *Tourism Management* 16(1): 49–59.

Hultsman, J. (1995) 'Just tourism: an ethical framework', *Annals of Tourism Research* 22(3): 553–567.

Hunter, C.J. (1995) 'On the need to re-conceptualise sustainable tourism development', *Journal of Sustainable Tourism* 3(3): 155–165.

Husbands, W. (1981) 'Centres, peripheries, tourism and socio-spatial development', *Ontario Geography* 17: 37–59.

Hvenegaard, G.T. (1994) 'Ecotourism: a status report and conceptual framework', *Journal of Tourism Studies* 5(2): 24–35.

Ingram, C.D. and Durst, P.B. (1987) 'Marketing nature-oriented tourism for rural development and wildlands management in developing countries: a bibliography', General Technical Report SE-44, Asheville, NC: US Dept of Agriculture, Forest Service, Southeastern Forest Experiment Station.

—— (1989) 'Nature-oriented tour operators: travel to developing countries', *Journal of Travel Research* 28(2): 11–15.

Inskeep, E. (1987) 'Environmental planning for tourism', *Annals of Tourism Research* 14(1): 118–135.

Iso-Ahola, S. (1982) 'Toward a social psychological theory of tourism motivation: a rejoinder', *Annals of Tourism Research* 9(2): 256–262.

Issaverdis, J.-P. (2001) 'The pursuit of excellence: benchmarking, accreditation, best practice and auditing', in D.B. Weaver (ed.) *The Encyclopedia of Ecotourism*, Wallingford, Oxon.: CAB International.

Iverson, T. (1997) 'Ecolabelling and tourism', TRINET communication, June 22.

Jansen-Verbeke, M. and Dietvorst, A. (1987) 'Leisure, recreation, tourism: a geographic view on integration', *Annals of Tourism Research* 14(3): 361–375.

Jenkins, C.L. (1991) 'Tourism development strategies', in L.J. Lickorish (ed.) *Developing Tourism Destinations*, Harlow: Longman.

—— (1994) 'Tourism in developing countries: the privatisation issue', in A.V. Seaton (ed.) *Tourism: The State of the Art*, Chichester: John Wiley.

Jenkins, J. and Wearing, S. (2003) 'Ecotourism and protected areas in Australia', in D.A. Fennell and R.K. Dowling (eds) *Ecotourism Policy and Planning*, Wallingford, Oxon.: CAB International.

Johnson, D.R. and Agee, J.K. (1988) 'Introduction to ecosystem management', in J.K. Agee and D.R. Johnson (eds) *Ecosystem Management for Parks and Wilderness*, Seattle: University of Washington Press.

Johst, D. (1982) 'Does wilderness designation increase recreation use?', unpublished report of the Bureau of Land Management, Washington, DC.

Jones, H. (1972) 'Gozo – the living showpiece', *Geographical Magazine* 45(1): 53–57.

Joppe, M. (1996) 'Sustainable community tourism development revisited', *Tourism Management* 17(7): 475–479.

Jordan, M. (2001) 'Brazil lays claims to the rain forest', *Wall Street Journal*, Eastern Edition, 31 August: A6, A11.

Jurowski, C. (1996) 'Tourism means more than money to the host community', *Parks and Recreation* 31(9): 110–118.

Jurowski, C., Muzaffer, U., Williams, D.R., and Noe, F.P. (1995) 'An examination of preferences and evaluations of visitors based on environmental attitudes: Biscayne Bay National Park', *Journal of Sustainable Tourism* 3(2): 73–86.

Kahle, L.R., Beatty, S.E., and Homer, P. (1986) 'Alternative measurement approaches to consumer values: the list of values (LOV) and values and life style (VALS)', *Journal of Consumer Research* 13(3): 404–409.

Karwacki, J. and Boyd, C. (1995) 'Ethics and ecotourism', *Business Ethics* 4(4): 225–232.

Kavallinis, I. and Pizam, A. (1994) 'The environmental impacts of tourism – whose responsibility is it anyway? The case study of Mykonos', *Journal of Travel Research* 33(2): 26–32.

Keller, C.P. (1987) 'Stages of peripheral tourism development – Canada's Northwest Territories', *Tourism Management* 8(1): 20–32.

Kellert, S.R. (1985) 'Birdwatching in American society', *Leisure Sciences* 7(3): 343–360.

—— (1987) 'The contributions of wildlife to human quality of life', in D.J. Decker and G.R. Goff (eds) *Valuing Wildlife: Economic and Social Perspectives*, Boulder: Westview.

Kelly, J.R. and Godbey, G. (1992) *The Sociology of Leisure*, State College, PA: Venture Publishing.

Kenchington, R.A. (1989) 'Tourism in the Galapagos Islands: the dilemma of conservation', *Environmental Conservation* 16(3): 227–236.

King, D.A. and Stewart, W.P. (1996) 'Ecotourism and commodification: protecting people and places', *Biodiversity and Conservation* 5(3): 293–305.

Knowles, M. (1980) *The Modern Practice of Adult Education*, New York: Cambridge University Press.

Kohlberg, L. (1981) *Essays on Moral Development*: Volume I, *The Philosophy of Moral Development*, New York: Harper & Row.

—— (1984) *Essays on Moral Development*: Volume II, *The Psychology of Moral Development*, New York: Harper & Row.

Kraus, R. (1997) *Recreation Programming: A Benefits-Driven Approach*, Toronto: Allyn & Bacon.

Kraus, R. and Allen, L. (1997) *Research and Evaluation in Recreation, Parks and Leisure Studies* (2nd. edn), Scottsdale, AZ: Gorsuch Scarisbrick, Publishers.

Kraus, R. and Curtis, J. (1990) *Creative Management in Recreation, Parks, and Leisure Services* (5th edn), St Louis, MO: Times Mirror/Mosby.

Kretchman, J.A. and Eagles, P.F.J. (1990) 'An analysis of the motives of ecotourists in comparison to the general Canadian population', *Society and Leisure* 13(2): 499–507.

Krippendorf, J. (1977) *Les dévoreurs des paysages*, Lausanne: 24 Heures.

—— (1982) 'Towards new tourism policies', *Tourism Management* 3: 135–148.

—— (1987) 'Ecological approach to tourism marketing', *Tourism Management* 8(2): 174–176.

Kusler, J.A. (1991) 'Ecotourism and resource conservation: introduction to issues', in J.A. Kusler (ed.) *Ecotourism and Resource Conservation: A Collection of Papers*, Volume 1. Madison, WI: Omnipress.

Kutay, K. (1989) 'The new ethic in adventure travel', *Buzzworm: The Environmental Journal* 1(4): 31–34.

LaFranchi, H. (1998) 'Bye-bye to Brazil's bio-paradise?', *Christian Science Monitor* 91(8): 12–14.

Laarman, J.G. and Durst, P.B. (1987) 'Nature travel and tropical forests', FPEI Working Paper Series, Southeastern Center for Forest Economics Research, North Carolina State University, Raleigh.

—— (1993) 'Nature tourism as a tool for economic development and conservation of natural resources', in J. Nenon and P.B. Durst (eds) *Nature Tourism in Asia: Opportunities and Constraints for Conservation and Economic Development*, Washington, DC: US Forest Service.

Laarman, J.G. and Gregersen, H. (1994) 'Making nature-based tourism contribute to sustainable development', *EPAT/MUCIA Policy Brief* 5: 1–6.

—— (1996) 'Pricing policy in nature-based tourism', *Tourism Management* 17(4): 247–254.

Langholz, J. and Brandon, K. (2001) 'Privately owned protected areas', in D.B. Weaver (ed.) *The Encyclopedia of Ecotourism*, Wallingford, Oxon.: CAB International Publishing.

Lawton, L. and Weaver, D.B. (2001). 'Ecotourism in modified spaces', in D.B. Weaver (ed.) *Encyclopedia of Ecotourism*, Wallingford, Oxon.: CAB International.

Leiper, N. (1981) 'Towards a cohesive curriculum in tourism: the case for a distinct discipline', *Annals of Tourism Research* 8(1): 69–84.

—— (1990) 'Tourist attraction systems', *Annals of Tourism Research* 17(3): 367–384.

Lemelin, H., McCarville, R., and Smale, B. (2002) 'The effects of context on price expectations for wildlife viewing opportunities in Churchill, Manitoba.' Report for the Manitoba Department of Conservation, and Parks Canada, Waterloo, Ontario.

León, C. and González, M. (1995) 'Managing the environment in tourism regions: the case of the Canary Islands', *European Environment* 5(6): 171–177.

Leslie, D. (1994) 'Sustainable tourism or developing sustainable approaches to lifestyle', *World Leisure and Recreation* 36(3): 30–36.

Lew, A. (1987) 'A framework of tourist attraction research', *Annals of Tourism Research* 14(4): 553–575.

Lickorish, L.J. (1991) 'Roles of government and the private sector', in L.J. Lickorish (ed.) *Developing Tourism Destinations*, Harlow: Longman.

Liddle, M. (1997) *Recreation Ecology: The Ecological Impact of Outdoor Recreation and Ecotourism*, London: Chapman and Hall.

Lindberg, K. (1991) *Policies for Maximising Nature Tourism's Ecological and Economic Benefits*, Washington, DC: World Resources Institute.

Lipske, K. (1992) 'How a monkey saved the jungle', *International Wildlife* 22(1): 38–43.

Liu, J.C. (1994) *Pacific Island Ecotourism: A Public Policy and Planning Guide*, Honolulu: Office of Territorial and International Affairs.

Loomis, L. and Graefe, A.R. (1992) 'Overview of NPCA's Visitor Impact Management Process', Paper presented at the Fourth World Congress on Parks and Protected Areas, Caracas, February 11–21.

Lothian, W.F. (1987) *A Brief History of Canada's National Parks*, Ottawa: Supply and Services Canada.

Lovejoy, T. (1992) 'Looking to the next millennium', *National Parks*, January/February: 41–44.

Lovelock, C.H. and Weinberg, C.B. (1984) *Marketing for Public and Nonprofit Managers*, Toronto: Wiley.

Lucas, R.C. (1964) 'Wilderness perception and use: the example of the Boundary Waters Canoe Area', *Natural Resources Journal* 3(3): 394–411.

Lucas, R.C. and Stankey, G.H. (1974) 'Social carrying capacity for backcountry recreation', in *Outdoor Recreation Research: Applying the Results*, USDA Forest Service General Technical Report NC-9, North Central Forest Experiment Station, St Paul, Minnesota, pp. 14–23.

Luhrman, D. (1997) 'WTO Manila Meeting', Internet communication, May 23, WTO Press and Communications.

Mabey, C. (1994) 'Youth leadership: commitment for what?' in S. York and D. Jordan, *Bold Ideas: Creative Approaches to the Challenge of Youth Programming*, Institute for Youth Leaders: University of Northern Iowa.

McArthur, S. (1997) 'Introducing the National Ecotourism Accreditation Program', *Australian Parks and Recreation* 33(2): 30–34.

Macbeth, J. (1994) 'To sustain is to nurture, to nourish, to tolerate and to carry on: can tourism? Trends in sustainable rural tourism development', *Parks and Recreation Magazine* 31(1): 42–45.

MacCannell, D. (1989) *The Tourist: A New Theory of the Leisure Class*, New York: Schocken Books.

McCarville, R. (1993) 'Keys to quality leisure programming', *Journal of Physical Education, Recreation and Dance*, October: 32–39.

McCool, S.F. (1985) 'Does wilderness designation lead to increased recreational use?' *Journal of Forestry*, January: 39–41.

—— (1995) 'Linking tourism, the environment, and concepts of sustainability: setting the stage', in S.F. McCool and A.E. Watson (eds) *Linking Tourism, the Environment, and Sustainability*, USDA Technical Report INT-GTR-323, Ogden, UT: US Department of Agriculture, Forest Service, Intermountain Research Station.

McFarlane, B.L. and Boxall, P.C. (1996) 'Participation in wildlife conservation by birdwatchers', *Human Dimensions of Wildlife* 1(3): 1–14.

McIntosh, C. (1992) 'Eco-tourism shows promise for the north', *Northern Ontario Business* 12(1): 9.

MacKay, K. and McIlraith, A. (1997) *Churchill Visitor Study: Seasonal Overview*, Winnipeg: Health, Leisure and Human Performance Research Institute, University of Manitoba.

McKenzie, J.F. and Smeltzer, J.L. (1997) *Planning, Implementing, and Evaluating Health Promotion Programs: A Primer* (2nd edn), Toronto: Allyn & Bacon.

McKercher, B. (1993a) 'The unrecognized threat to tourism: can tourism survive "sustainability"?', *Tourism Management* 14(2): 131–136.

—— (1993b) 'Some fundamental truths about tourism: understanding tourism's social and environmental impacts', *Journal of Sustainable Tourism* 1(1): 6–16.

Mackie, J.L. (1977) *Ethics: Inventing Right and Wrong*, Harmondsworth: Penguin Books.

MacKinnon, B. (1995) 'Beauty and the beasts of ecotourism', *Business Mexico* 5(4): 44–47.

McNeely, J.A. (1988) *Economics and Biological Diversity*, Gland, Switzerland: International Union for the Conservation of Nature and Natural Resources.

—— (1993) 'People and protected areas: partners in prosperity', in E. Kemf (ed.) *The Law of the Mother: Protecting Indigenous People in Protected Areas*, San Francisco: Sierra Club.

Madrigal, R. (1995) 'Personal values, traveler personality type, and leisure travel style', *Journal of Leisure Research* 27(2): 125–142.

Madrigal, R. and Kahle, L.R. (1994) 'Predicting vacation activity preferences on the basis of value-system segmentation', *Journal of Travel Research* 32(3): 22–28.

Mahoney, E.M. (1988) 'Recreation and tourism marketing', unpublished paper, Michigan State University, Ann Arbor, Michigan.

Malloy, D.C. and Fennell, D.A. (1998a) 'Ecotourism and ethics: moral development and organizational cultures', *Journal of Travel Research* 36(4): 47–56.

—— (1998b) 'Codes of ethics and tourism: an explanatory content analysis', *Tourism Management*.

Manning, T. (1996) 'Tourism: where are the limits?', *Ecodecision* 20(Spring): 35–39.

Marinelli, L. (1997) 'Ecotourists take to the Hawaiian hills', *The Toronto Star*, October 25.

Maslow, A. (1954) *Motivation and Personality*, New York: Harper & Row.

Mason, J.G. (1960) *How to be a More Effective Executive*, New York: McGraw-Hill.

Mason, P. (1997) 'Ecolabelling and tourism', TRINET communication, June 22.

Mason, P. and Mowforth, M. (1995) *Codes of Conduct in Tourism*, University of Plymouth, Department of Geographical Sciences Occasional Paper No. 1.

Mathieson, A. and Wall, G. (1982) *Tourism: Economic, Physical, and Social Impacts*, London: Longman.

Mayo, E.F. (1975) 'Tourism and the national parks: a psychographic and attitudinal study', *Journal of Leisure Research* 14(1): 14–18.

Mayur, R. (ed.) (1996) *Earth, Man and Future*, Mumbai, India: International Institute for Sustainable Future.

Menkhaus, S. and Lober, D.J. (1996) 'International ecotourism and the valuation of tropical rainforests in Costa Rica', *Journal of Environmental Management* 47: 1–10.

Merry, W. (1994) *St. John Ambulance Official Wilderness First-Aid Guide*, Toronto: McClelland & Stewart Inc.

Metelka, C.J. (1981) *The Dictionary of Tourism*, Urbana–Champaign, IL: Merton House Co.

—— (1990) *The Dictionary of Hospitality, Travel and Tourism*, Albany, NY: Delmar Publishers.

Meyer-Arendt, K.J. (1985) 'The Grand Isle, Louisiana resort cycle', *Annals of Tourism Research* 12: 449–465.

Mieczkowski, Z.T. (1981) 'Some notes on the geography of tourism: a comment', *Canadian Geographer* 25: 186–191.

Milgrath, L. (1989) 'An inquiry into values for a sustainable society: a personal statement', in L. Milgrath (ed.) *Envisioning a Sustainable Society*, Albany, NY: SUNY Press.

Mill, R.C. and Morrison, A.M. (1985) *The Tourism System*, Englewood Cliffs, NJ: Prentice-Hall.

Miller, B.R. (1996) 'The global structure of the nature tourism industry: ecotourists, tour operators, and local businesses', *Journal of Travel Research* 35(2): 11–18.

Miller, K.R. (1976) 'Global dimensions of wildlife management in relation to development and environmental conservation in Latin America', Proceedings of Regional Expert Consultation on Environment and Development, Bogotá, July 5–10, 1976. Santiago, Chile: Food and Agriculture Organization.

—— (1989) *Planning National Parks for Ecodevelopment: Methods and Cases from Latin America*, Washington, DC: Peace Corps.

Miller, M.L. and Kaye, B.C. (1993) 'Coastal and marine ecotourism: a formula for sustainable development?', *Trends* 30(2): 35–41.

Milne, S. and Grekin, J. (1992) 'Travel agents as information brokers: the case of the Baffin Region, Northwest Territories', *The Operational Geographer* 10(3): 11–14.

Mintzberg, H. (1996) 'The myth of "Society Inc."', *The Globe and Mail Report on Business Magazine*, October: 113–117.

Mirabella, M. (1998) 'Brazil hopes new push for ecotourism can help save Amazon rain forest', (accessed 15 May 2002: www.cnn.com/travel/news/9801/28/amazon.ecotourism).

Mitchell, B. (1989) *Geography and Resource Analysis*, New York: Longman.

—— (1994) 'Sustainable development at the village level in Bali, Indonesia', *Human Ecology* 22(2): 189–211.

Mitchell, G.E. (1992) *Ecotourism Guiding: How to Start your Career as an Ecotourism Guide*, n.p.: G.E. Mitchell.

Mitchell, L.S. (1984) 'Tourism research in the United States: a geographical perspective', *Geojournal* 9(1): 5–15.

Mlinaric, I.B. (1985) 'Tourism and the environment: a case for Mediterranean cooperation', *International Journal of Environmental Studies* 25: 239–245.

Moore, S. and Carter, B. (1993) 'Ecotourism in the 21st century', *Tourism Management* 14(2): 123–130.

Morrison, A.M., Hsieh, S., and Wang, C.-Y. (1992) 'Certification in the travel and tourism industry', *Journal of Tourism Studies* 3(2): 32–40.

Moscardo, G., Morrison, A.M., and Pearce, P.L. (1996) 'Specialist accommodation and ecologically-sustainable tourism', *Journal of Sustainable Tourism* 4(1): 29–54.

Mountjoy, A.B. (1971) *Developing the Underdeveloped Countries*, New York: John Wiley.

Munasinghe, M. (1994) 'Economic and policy issues in natural habitats and protected areas', in M. Munasinghe and J. McNeely (eds) *Protected Area Economics and Policy: Linking Conservation and Sustainable Development*, Washington, DC: The World Bank.

Munt, I. and Higinio, E. (1993) 'Belize – eco-tourism gone awry', *In Focus* 9: 18–19.

Murphy, P.E. (1983) 'Tourism as a community industry: an ecological model of tourism development', *Tourism Management* 4(3): 180–193.

Myers, N. (1980) *The Sinking Ark*, Oxford: Pergamon Press.

Nash, R. (1982) *Wilderness and the American Mind*, New Haven: Yale University Press.

Naylon, J. (1967) 'Tourism – Spain's most important industry', *Geography* 52: 23–40.

Nelson, J.G. (1991) 'Sustainable development, conservation strategies, and heritage', in B. Mitchell (ed.) *Resource Management and Development*, Oxford: Oxford University Press.

—— (1993) 'Planning and managing national parks and protected areas: a human ecological approach', Paper presented at the Ecosystem Management for Managers Workshop, University of Waterloo, Waterloo, Ontario, January 18–21.

—— (1994). 'The spread of ecotourism: some planning implications', *Environmental Conservation* 21(1): 248–255.

Nelson, J.G., Butler, R., and Wall, G. (1993) *Tourism and Sustainable Development: Monitoring, Planning, Managing*, Waterloo, Ontario: Department of Geography Publication Series No. 37, University of Waterloo.

Newsome, D., Moore, S.A., and Dowling, R.K. (2002) *Natural Area Tourism: Ecology, Impacts, and Management*, Clevedon, Somerset: Channel View Publications.

Norris, R. (1992) 'Can ecotourism save natural areas?', *Parks*, January/February: 31–34.

Notzke, C. (1994) 'Aboriginal people and protected areas', Paper presented at Saskatchewan's Protected Areas Conference, University of Regina, Regina, Saskatchewan, June.

O'Gara, G. (1996) 'A natural history of the Yellowstone tourist', *Sierra* 81(2): 54–59, 83–85.

Olesen, R.M. and Schettini, P. (1994) 'From classroom to cornice: training the adventure tourism professional', in A.V. Seaton (ed.) *Tourism: The State of the Art*, Chichester: John Wiley.

Ontario Ministry of Tourism and Recreation (1985a) *Planning Recreation: A Manual of Principles and Practices*, Toronto: Government of Ontario.

Ontario Ministry of Tourism and Recreation (1985b) *Better Planning for Better Recreation*, Toronto: Government of Ontario.

Oppermann, M. (1997) 'Ecolabelling and tourism', TRINET communication, June 22.

Orams, M.B. (1995) 'Towards a more desirable form of ecotourism', *Tourism Management* 16(1): 3–8.

O'Reilly, A.M. (1986) 'Tourism carrying capacity: concepts and issues', *Tourism Management* 7(4): 254–258.

Ortolano, L. (1984) *Environmental Planning and Decision Making*, New York: John Wiley.

Page, S. and Dowling, R. (2002) *Ecotourism*, Essex, UK: Person Education.

Passmore, J. (1974) *Man's Responsibility for Nature*, New York: Scribner.

Payne, D. and Dimanche, F. (1996) 'Towards a code of conduct for the tourism industry: an ethics model', *Journal of Business Ethics* 15: 997–1007.

Payne, R.J. and Graham, R. (1993) 'Visitor planning and management in parks and protected areas', in P. Dearden and R. Rollins (eds) *Parks and Protected Areas in Canada*, Toronto: Oxford University Press.

Pearce, D.G. (1985) 'Tourism and environmental research: a review', *International Journal of Environmental Studies* 25: 247–255.

—— (1991) *Tourist Development*, London: Longman.

Pearce, P.L. (1982) *The Social Psychology of Tourist Behaviour*, Oxford: Pergamon Press.

Peattie, K. (1995) *Environmental Marketing Management: Meeting the Green Challenge*, London: Pitman.

Peterson, G. (1996) 'Four corners of human ecology: different paradigms of human relationships with the earth', in B.L. Driver, D. Dustin, T. Baltic, G. Elsner, and G. Peterson (eds) *Nature and the Human Spirit: Toward an Expanded Land Management Ethic*, State College, PA: Venture.

Pfafflin, G.F. (1987) 'Concern for tourism: European perspective and response', *Annals of Tourism Research* 14(4): 576–579.

Philipsen, J. (1995) 'Nature-based tourism and recreation: environmental change, perception, ideology and practice', in G.J. Ashworth and A.G.J. Dietvorst (eds) *Tourism and Spatial Transformations: Implications for Policy and Planning*, Wallingford, Oxon.: CAB International.

Phillips, A. (1985) 'Socio-economic development in the "national parks" of England and Wales', *Parks* 10(1): 1–5.

Pinchot, G. (1910) *The Fight for Conservation*, New York: Harcourt Brace.

—— (1947) *Breaking New Ground*, Seattle: University of Washington Press.

Pinhey, T.K. and Grimes, M.D. (1979) 'Outdoor recreation and environmental concern: a re-examination of the Dunlap–Heffernan thesis', *Leisure Sciences* 2: 1–11.

Pitt, D.G. and Zube, E.H. (1987) 'Management of natural environments', *Handbook of Environmental Psychology* 1: 1009–1041.

Pollock, N.C. (1971) 'Serengeti', *Geography* 56(2): 145–147.

Preece, N., van Oosterzee, P., and James, D. (1995) *Biodiversity Conservation and Ecotourism: An Investigation of Linkages, Mutual Benefits and Future Opportunities*, Canberra: Department of the Environment, Sport, and Territories.

Pretty, J. and Pimbert, M. (1995) 'Trouble in the Garden of Eden', *Guardian* (London), May 13, section D.

Price, A.R.G. and Firaq, I. (1996) 'The environmental status of reefs on Maldivian resort islands: a preliminary assessment for tourism planning', *Aquatic Conservation: Marine and Freshwater Ecosystems* 6(2): 93–106.

Priest, S. (1990) 'The semantics of adventure education', in J.C. Miles and S. Priest (eds) *Adventure Education*, State College, PA: Venture Publishing.

Quinn, B. (1990) 'The essence of adventure', in J.C. Miles and S. Priest (eds) *Adventure Education*, State College, PA: Venture Publishing.

Rajotte, F. (1980) 'Tourism in the Pacific', in F. Rajotte and R. Crocombe (eds) *Pacific Tourism – As Islanders See It*, Suva, Fiji: Institute of Pacific Studies, University of the South Pacific, pp. 1–14.

Redclift, M. (1987) *Sustainable Development: Exploring the Contradictions*, London: Methuen.

Reid, R., Stone, M., and Whiteley, T. (1995) *Economic Value of Wilderness Protection and Recreation in British Columbia*, FRDA II.

Reidenbach, R.E. and Robin, D.P. (1988) 'Some initial steps toward improving the measurement of ethical evaluations of marketing activities', *Journal of Business Ethics* 7: 871–879.

—— (1990) 'Toward the development of a multidimensional scale for improving evaluations of business ethics', *Journal of Business Ethics* 9: 639–653.

Reingold, L. (1993) 'Identifying the elusive ecotourist', in *Going Green*, a supplement to *Tour and Travel News*, October 25, pp. 36–37.

Relph, E. (1976) *Place and Placelessness*, New York: Methuen.

Reynolds, L. (1992) 'Montserrat to target "ecotourists"', *The Globe and Mail*, February 22. Travel Section.

Reynolds, P.C. and Braithwaite, D. (2001) 'Towards a conceptual framework for wildlife tourism', *Tourism Management* 22: 31–42.

Richter, L.K. (1991) 'Political issues in tourism policy: a forecast', in D.E. Hawkins and J.R.B. Ritchie (eds), *World Travel and Tourism Review* Vol. 1, London: CAB International: 189–193.

Rivers, P. (1973) 'Tourist troubles', *New Society* 23: 250.

Rollins, R. (1993) 'Managing the national parks', in P. Dearden and R. Rollins (eds) *Parks and Protected Areas in Canada*, Toronto: Oxford University Press.

Rollins, R. and Dearden, P. (1993) 'Challenges for the future', in P. Dearden and R. Rollins (eds) *Parks and Protected Areas in Canada*, Toronto: Oxford University Press.

Romeril, M. (1985) 'Tourism and the environment – towards a symbiotic relationship', *International Journal of Environmental Studies* 25: 215–218.

Romero, S. (2000) 'Amazon seeks to green economy', *The Toronto Star*, 19 June, section E6.

Russell, D., Bottrill, C., and Meredith, G. (1995) 'International ecolodge survey', in *The Ecolodge Sourcebook – for Planners and Managers*, North Bennington, VT: The Ecotourism Society.

Russell, R.V. (1982) *Planning Programs in Recreation*, St Louis, MO: Mosby.

Ryan, C. (1991) *Recreational Tourism: A Social Science Perspective*, New York: Routledge.

—— (1997) 'Ecolabelling and tourism', TRINET communication, June 22.

Ryel, R. and Grasse, T. (1991) 'Marketing ecotourism: attracting the elusive ecotourist', in T. Whelan (ed.) *Nature Tourism: Managing for the Environment*, Washington, DC: Island Press.

Sadler, B. (1989) 'National parks, wilderness preservation, and native peoples in northern Canada', *Natural Resources Journal* 29: 185–204.

—— (1992) 'Introduction', in S. Hawkes and P. Williams (eds) *The Greening of Tourism: From Principles to Practice*, Centre for Tourism Policy and Research, Simon Fraser University, British Columbia.

Scace, R.C., Grifone, E., and Usher, R. (1992) *Ecotourism in Canada*, Consulting report prepared for the Canadian Environmental Advisory Council, Hull, Quebec: Minister of Supply and Services.

Scalet, C.G. and Adelman, I.R. (1995) 'Accreditation of fisheries and wildlife programs', *Fisheries* 20(2): 8–13.

Schein, E.H. (1985) *Organizational Culture and Leadership*, San Francisco: Jossey-Bass.

Schermerhorn, J.R. (1996) *Management* (5th edn), New York: John Wiley & Sons.

Seale, R.G. (1992) 'Aboriginal societies, tourism and conservation: the case of Canada's Northwest Territories', Paper presented at the Fourth World Congress on Parks and protected Areas, Caracas, February 10–21.

Searle, M.S. and Brayley, R.E. (1993) *Leisure Services in Canada*, State College, PA: Venture Publishing, Inc.

Sessoms, H.D. (1991) 'Certifying park and recreation: the American experience', *Recreation Canada*, July: 21–23.

Shackleford, P. (1985) 'The World Tourism Organisation – 30 years of commitment to environmental protection', *International Journal of Environmental Studies* 25: 257–263.

Shelby, B. and Heberlein, T. (1986) *Carrying Capacity in Recreational Settings*, Oregon: Oregon State University Press.

Shelby, B. and Vaske, J.J. (1991) 'Using normative data to develop evaluative standards for resource management: a comment on three recent papers', *Journal of Leisure Research* 23(2): 173–187.

Sherman, P.B. and Dixon, J.A. (1991) 'The economics of nature tourism: determining if it pays', in T. Whelan (ed.) *Nature Tourism: Managing for the Environment*, Washington, DC: Island Press.

Shields, R. (1991) *Places on the Margin*, New York: Routledge.

Shores, J.N. (1992) 'The challenge of ecotourism: a call for higher standards', Paper presented at the Fourth World Congress on Parks and Protected Areas, Caracas, February 10–21.

Short, J.R. (1991) *Imagined Country*, New York: Routledge, Chapman & Hall.

Shundich, S. (1996) 'Ecotourists: dollars, sense and the environment', *Hotels*, March: 34–40.

Silverberg, K.E., Backman, S.J., and Backman, K.F. (1996) 'A preliminary investigation into the psychographics of nature-based travelers to the Southeastern United States', *Journal of Travel Research* 35(2): 19–28.

Silverman, G.S. (1992) 'Accrediting undergraduate programs in environmental health science and protection', *The Environmental Professional* 14(4): 319–24.

Simpson, J. (1983) 'The discovery of Shetland from *The Pirate* to the tourist board', in *Shetland and the Outside World 1469–1969*, New York: Oxford University Press.

Simpson, K. (2001) 'Strategic planning and community involvement as contributors to sustainable tourism development', *Current Issues in Tourism* 4(1): 3–41.

Sims, R.R. (1991) 'The institutionalization of organizational ethics', *Journal of Business Ethics* 10: 493–506.

Singer, M.G. (1986) 'Ethics, science and moral philosophy', in J.P. DeMarco and R.M. Fox (eds) *New Directions in Ethics: The Challenge of Applied Ethics*, New York: Routledge & Kegan Paul.

Slocombe, D.S. and Nelson, J.G. (1992) 'Management issues in hinterland national parks: a human ecological approach', *Natural Areas Journal* 12(4): 206–215.

Smale, B.J.A. and Reid, D.G. (1995) 'Public policy on recreation and leisure in urban Canada', in R. Loreto and T. Price (eds) *Urban Policy Issues: Canadian Perspectives*, 2nd edn, Toronto: Oxford University Press.

Smith, S.L.J. (1990a) *Dictionary of Concepts in Recreation and Leisure Studies*, New York: Greenwood Press.

—— (1990b) *Tourism Analysis*, New York: Longman.

Snell, M.B. (2001) 'Gorillas in the crossfire', *Sierra* 86(6): 30–34.

Sofield, T. and Li, F.M.S. (2003) 'Processes in formulating an ecotourism policy for nature reserves in Yunnan Province, China', in D.A. Fennell and R.K. Dowling (eds) *Ecotourism Policy and Planning*, Wallingford, Oxon.: CAB International.

Sproule, K.W. (1996) 'Community-based ecotourism development: identifying partners in the process', Paper presented at the Ecotourism Equation: Measuring the Impacts (ISTF) Conference, Yale School of Forestry and Environmental Studies, April 12–14.

Stanford, C.B. (1999) 'Gorilla warfare', *The Sciences* (July/August): 18–23.

Stankey, G.H. and McCool, S.F. (1984) 'Carrying capacity in recreational settings: evolution, appraisal, and application', *Leisure Sciences* 6(4): 453–473.

Steele, P. (1993) 'The economics of eco-tourism', *In Focus* 9: 7–9.

—— (1995) 'Ecotourism: an economic analysis', *Journal of Sustainable Tourism* 3(1): 29–44.

Stevens, B. (1994) 'An analysis of corporate ethical code studies: "Where do we go from here?"', *Journal of Business Ethics* 13: 63–69.

Stonehouse, B. (2001) 'Polar environments', in D.B. Weaver (ed.) *The Encyclopedia of Ecotourism*, Wallingford, Oxon.: CAB International.

Swart, S. and Saayman, M. (1997) 'Legislative restrictions on the tourism industry: a South African perspective', *World Leisure and Recreation* 39(1): 24–30.

Tangley, L. (1988) 'Who's polluting Antarctica?', *BioScience* 38(9): 590–594.

Texas Parks and Wildlife (1996) *Nature Tourism in the Lone Star State*, Austin: Texas Parks and Wildlife Development.

Theophile, K. (1995) 'The forest as a business: is ecotourism the answer?', *Journal of Forestry* 93(3): 25–27.

Thomlinson, E. and Getz, D. (1996) 'The question of scale in ecotourism: case study of two small ecotour operators in the Mundo Maya region of Central America', *Journal of Sustainable Tourism* 4(4): 183–200.

Thompson, P. (1995) 'The errant e-word: putting ecotourism back on track', *Explore*, 73: 67–72.

Tibbetts, J. (1995–6) 'A walk on the wild side', *Coastal Heritage* 10(3): 3–9.

Tims, D. (1996) 'The perspective of outfitters and guides', in B.L. Driver, D. Dustin, T. Baltic, G. Elsner, and G. Peterson (eds) *Nature and the Human Spirit: Toward an Expanded Land Management Ethic*, State College, PA: Venture.

Tisdell, C. (1995) 'Investment in ecotourism: assessing its economics', *Tourism Economics* 1(4): 375–387.

Tompkins, L. (1996) *A Description of Wilderness Tourism and Outfitting in the Yukon*, Whitehorse: Department of Tourism.

Tourism Concern (1992) *Beyond the Green Horizon: Principles for Sustainable Tourism*, United Kingdom: World Wildlife Fund.

Tourism Industry Association of Canada (1995) *Code of Ethics and Guidelines for Sustainable Tourism*, Ottawa: Tourism Industry Association of Canada.

Travis, A.S. (1982) 'Physical impacts: trends affecting tourism', *Tourism Management* 3: 256–262.

Tribe, J., Font, X., Griffiths, N., Vickery, R., and Yale, K. (2000) *Environmental Management for Rural Tourism and Recreation*, London: Cassell.

Tuan, Y.-F. (1971) 'Geography, phenomenology, and the study of human nature', *The Canadian Geographer* 14: 193–201.

Turnbull, C. (1981) 'East African safari', *Natural History* 90(5): 26–34.

Uhlik, K.S. (1995) 'Partnership step by step: a practical model of partnership formation', *Journal of Park and Recreation Administration* 13(4): 13–24.

UNEP/WTO (2002) Quebec Declaration on Ecotourism: www.uneptie.org/pc/tourism/documents/ecotourism/WESoutcomes/Quebec-Declar-eng.pdf (visited 31 August 2002).

United Nations Environment Programme Industry and Environment (1995) *Environmental Codes of Conduct for Tourism*, Technical Report No. 29, Paris: UNEP.

Upchurch, R.S. and Ruhland, S.K. (1995) 'An analysis of ethical work climate and leadership relationship in lodging operations', *Journal of Travel Research* 34(2): 36–42.

Urry, J. (1992) 'The tourist gaze and the "environment"', *Theory, Culture and Society* 9: 1–26.

US Department of the Interior (1993) *Guiding Principles of Sustainable Design*, Denver, CO: Denver Service Center.

Valentine, P.S. (1993) 'Ecotourism and nature conservation: a definition with some recent developments in Micronesia', *Tourism Management* 14(2): 107–115.

van der Merwe, C. (1996) 'How it all began: the man who "coined" ecotourism tells us what it means', *African Wildlife* 50(3): 7–8.

van Liere, K.D. and Noe, F.P. (1981) 'Outdoor recreation and environmental attitudes: further examination of the Dunlap–Heffernan thesis', *Rural Sociology* 46: 501–513.

Veal, A.J. (1992) *Research Methods for Leisure and Tourism: A Practical Guide*, Harlow: Longman Group.

Veverka, J.A. (1994) 'Interpretation as a management tool', *Environmental Interpretation* 9(2): 18–19.

Vogeler, I. and DeSouza, A. (1980) *Dialectics of Third World Development*, New Jersey: Allanheld, Osmun.

Wagar, J.A. (1964) 'The carrying capacity of wildlands for recreation', Society of American Foresters, Forest Service Monograph 7: 23.

Wall, G. (1982) 'Cycles and capacity: incipient theory or conceptual contradiction', *Tourism Management* 3(3): 188–192.

—— (1993) 'International collaboration in the search for sustainable tourism in Bali, Indonesia', *Journal of Sustainable Tourism* 1(1): 38–47.

—— (1994) 'Ecotourism: old wine in new bottles?', *Trends* 31(2): 4–9.

Wall, G. and Wright, C. (1977) *The Environmental Impact of Outdoor Recreation*, Publication Series No. 11. Department of Geography, University of Waterloo, Ontario.

Wallace, G.N. (1993) 'Wildlands and ecotourism in Latin America', *Journal of Forestry* 91(2): 37–40.

Wallace, G.N. and Pierce, S.M. (1996) 'An evaluation of ecotourism in Amazonas, Brazil', *Annals of Tourism Research* 23(4): 843–873.

Walle, A.H. (1995) 'Business ethics and tourism: from micro to macro perspectives', *Tourism Management* 16(4): 263–268.

Watson, R. (1996) 'Risk management: a plan for safer activities', *CAHPERD Journal*, Spring: 13–17.

Wearing, S. (1994) 'Social and cultural perspectives in training for indigenous ecotourism development'. Unpublished paper.

—— (1995) 'Professionalisation and accreditation of ecotourism', *Leisure and Recreation* 37(4): 31–36.

Wearing, S. and Neil, J. (1999) *Ecotourism: Impacts, Potential and Possibilities*, Oxford: Butterworth & Heinemann.

Weaver, D.B. (1990) 'Grand Cayman Island and the resort cycle concept', *Journal of Travel Research* 29(2): 9–15.

—— (1991) 'Alternative to mass tourism in Dominica', *Annals of Tourism Research* 18: 414–432.

—— (1993) 'Ecotourism in the small island Caribbean', *GeoJournal* 31: 457–465.

—— (1995) 'Alternative tourism in Monserrat', *Tourism Management* 16(8): 593–604.

—— (1998) *Ecotourism in the Less Developed World*, London: CAB International.

—— (2001a) *Ecotourism*, Milton, Queensland: John Wiley & Sons.

—— (2001b) *The Encyclopedia of Ecotourism*, Wallingford, Oxon.: CAB International.

Weaver, D.B. and Schluter, R. (2001) 'Latin America and the Caribbean', in D.B. Weaver (ed.) *The Encyclopedia of Ecotourism*, Wallingford, Oxon.: CAB International.

Weaver, G. and Wishard-Lambert, V. (1996) 'Community tourism development: an opportunity for park and recreation departments', *Parks and Recreation* 31(9): 78–83.

Weiler, B. (1993) 'Nature-based tour operators: are they environmentally friendly or are they faking it?', *Tourism Recreation Research* 18(1): 55–60.

Weiler, B. and Davis, D. (1993) 'An exploratory investigation into the roles of the nature-based tour leader', *Tourism Management* 14(2): 91–98.

Weiler, B. and Ham, S. (2001) 'Tour guides and interpretation', in D.B. Weaver (ed.) *The Encyclopedia of Ecotourism*, Wallingford, Oxon.: CAB International.

Weschler, I.R. (1962) *Issues in Human Relations Training*, Washington, DC: National Training Laboratories.

Western, D. (1993) 'Defining ecotourism', in K. Lindberg and D.E. Hawkins (eds) *Ecotourism: A Guide for Planners and Managers*, North Bennington, VT: The Ecotourism Society.

Western, D. and Thresher, P. (1973) *Development Plans for Amboseli*, World Bank Report, Nairobi.

Wheeler, M. (1994) 'The emergence of ethics in tourism and hospitality', *Progress in Tourism, Recreation, and Hospitality Management* 6: 46–56.

Wheeller, B. (1994) 'Egotourism, sustainable tourism and the environment – a symbiotic, symbolic or shambolic relationship', in A.V. Seaton (ed.) *Tourism: The State of the Art*, Chichester: John Wiley.

White, D. (1993) 'Tourism as economic development for native people living in the shadow of a protected area: a North American case study', *Society and Natural Resources* 6: 339–345.

White, L. Jr (1971) 'The historic roots of our ecologic crisis', in R.M. Irving and G.B. Priddle (eds) *Crisis*, London: Macmillan.

Wight, P.A. (1993a) 'Sustainable ecotourism: balancing economic, environmental and social goals within an ethical framework', *Journal of Tourism Studies* 4(2): 54–66.

—— (1993b) 'Ecotourism: ethics or eco-sell?', *Journal of Travel Research* 21(3): 3–9.

—— (1995) 'Greening of remote tourism lodges', Paper presented at Shaping Tomorrow's North: The Role of Tourism and Recreation, Lakehead University, Thunder Bay, Ontario, October 12–15.

—— (1996) North American ecotourists: market profile and trip characteristics', *Journal of Travel Research* 34(4): 2–10.

Wilensky, H.L. (1964) 'The professionalization of everyone?', *American Journal of Sociology* 70(2): 137–158.

Wilkinson, P. (1991) 'Travel agents as information brokers: the cases of Anguilla and Dominica', *Operational Geographer* 9(3): 37–41.

Williacy, S. and Eagles, P.F.J. (1990) *An Analysis of the Federation of Ontario Naturalists' Canadian Nature Tours Programme*, Department of Recreation and Leisure Studies, University of Waterloo, Waterloo, Ontario.

Williams, P.W. (1992) 'A local framework for ecotourism development', *Western Wildlands* 18(3): 14–19.

Wilson, A. (1992) *The Culture of Nature*, Cambridge, MA: Blackwell.

Wilson, M. (1987) 'Nature oriented tourism in Ecuador: assessment of industry structure and development needs', (FPEI) North Carolina State University, Raleigh, North Carolina, No. 20.

Windsor, R., Baranowski, T., Clark, N., and Cutter, G. (1994) *Evaluation of Health Promotion, Health Education, and Disease Prevention Programs*, Mountain View, CA: Mayfeld.

Winpenny, J.T. (1982) 'Issues in the identification and appraisals of tourism projects in developing countries', *Tourism Management* 3(4): 218–221.

World Commission on Environment and Development (1987) *Our Common Future*, Oxford: Oxford University Press.

Worthen, B.R., Sanders, J.R., and Fitzpatrick, J.L. (1997) *Program Evaluation: Alternative Approaches and Practical Guidelines*, New York: Longman.

Wright, J.R. (1983) *Urban Parks in Ontario Part I: Origins to 1860*, Toronto: Ministry of Tourism and Recreation.

—— (1987) 'The university and the recreation profession', *Recreation Canada* 45(3): 14–18.

Wyatt, S. (1997) 'Dialogue, reflection, and community', *The Journal of Experiential Education* 20(2): 80–85.

Yee, J.G. (1992) *Ecotourism Market Survey: A Survey of North American Ecotourism Tour Operators*, San Francisco: PATA.

Young, B. (1983) 'Touristization of a traditional Maltese fishing–farming village: a general model', *Tourism Management* 4(1): 35–41.

Young, F.J.L. (1964) 'The contracting out of work: Canadian and U.S.A. industrial relations experience', Industrial Relations Centre, Queen's University, Kingston, Ontario.

Ziffer, K. (1989) *Ecotourism: The Uneasy Alliance*, Working Paper No. 1, Conservation International, Washington, DC.

Zimmerman, E.W. (1951) *World Resources and Industries*, New York: Harper.

Zurick, D.N. (1992) 'Adventure travel and sustainable tourism in the peripheral economy of Nepal', *Annals of the Association of Geographers* 82(4): 608–628.

# Index

Note: Names of places are listed under their country. References to figures, tables and plates are given as e.g. 21*f*, 25*t*, **69.**